Glass, Brass, and Chrome

Glass, Brass, & Chrome

THE AMERICAN 35mm MINIATURE CAMERA

by KALTON C. LAHUE

and JOE A. BAILEY

University of Oklahoma Press
Norman

Books by Kalton C. Lahue

Continued Next Week: A History of the Moving Picture Serial (Norman, 1964)
World of Laughter: The Motion Picture Comedy Short, 1910–1930 (Norman, 1966)
Kops and Custards: The Legend of Keystone Films (with Terry Brewer, Norman, 1968)
Bound and Gagged (Cranbury, N.J., 1968)
Collecting Classic Films (New York, 1970)
Clown Princes and Court Jesters (with Samuel Gill, Cranbury, N.J., 1970)
Winners of the West (Cranbury, N.J., 1970)
Dreams for Sale (Cranbury, N.J., 1971)
Ladies in Distress (Cranbury, N.J., 1971)
Mack Sennett's Keystone (Cranbury, N.J., 1971)
Collecting Vintage Cameras (with Joe A. Bailey, New York, 1971)
Glass, Brass, and Chrome: The American 35mm Miniature Camera (with Joe A. Bailey, Norman, 1972)

Library of Congress Catalog Card Number: 74–160496

ISBN: 978-0-8061-3434-5 (paper)

For Paul Bond, Kevin, and Kory

Preface

The age of the American 35mm camera coincided with three tumultuous decades in American history. Born in the depression years of the 1930's, grown to maturity during World War II, the American 35mm had reached its dotage by the early fifties. If you are a photo hobbyist today, or if you were one during the "golden years" of the American 35mm, the story you are about to read concerns an energetic and innovative America and an era which proved to be a good time for a photographer to be alive. In the span of just over thirty years the American photographic industry converted what had been a rich man's toy into an everyman's gadget. It created a new subeconomy in the industry, made candid photography a household term, and finally innovated itself out of the market it had created, returning the implementation of the idea to its foreign origin.

There is something about a group of fine mechanical parts assembled into pleasing shapes with skillful craftsmanship that creates a desire for ownership and a pride of possession, whether it be an automobile, a mobile sculpture, or a camera. Not that all American 35mm cameras were works of art; indeed, many models were quite the opposite. It is significant, however, that those cameras which became popular had an appeal to their buyers in terms other than their function of taking pictures. During our research for this volume, we had frequent debates with photographic historians on whether our work would be complete if it accented the hardware and simply assumed that the camera purchasers achieved, at least to their own satisfaction, artistically pleasing results. Since the choice was ours, we allowed our side to win. Photography as an art form has been evaluated by those better qualified and, in truth, more interested in art than we are. Our attempt is a history of American 35mm hardware, its evolution from a German idea, and the role it played in making photography America's number-one hobby today.

We acknowledge that the purist will contend that we have incorrectly appropriated the term *miniature* and corrupted it for our own use. Before the advent of the American 35mm camera and for some time after, *miniature* was used by everyone to describe any camera taking pictures 2¼ × 3¼ inches and smaller. But by the late thirties *miniature* had become closely associated with the 35mm camera, and with the demise of the folding camera and small press cameras following World War II and the advent of the subminiature cameras (smaller than 35mm), the term became accepted as belonging to the 35mm, and it is in this sense that we have used it.

Unfortunately, some of the companies which produced well-known cameras of the era were small family-owned organizations and did not divulge production and other data. Nor did the serial numbers of the cameras indicate anything specific to the outsider. With the passing of years and the deaths of many involved, some portions of a fascinating story have become almost impossible to assemble.

For these reasons we admit that the record is still an incomplete one. But we have attempted to record an era of creativity and innovation which holds a deep interest for us, and we found in our research that a growing number of other camera fans seem to feel the same way.

In reporting on many specific cameras, we have been forced to make judgments which have not before appeared in print, but we have done so on the basis of firsthand knowledge, either of our own or of persons with recognized technical expertise. While respecting the cameras and the companies involved for what they were, we have tried to present an objective picture both of the industry and of the products of an exciting photographic age.

No work of this nature could be accomplished without the help of many interested people, and while space does not permit acknowledging all those who graciously lent their time, shared their memories, and provided cameras which the authors do not have in their collections, the following deserve special mention: W. F. Armstrong, C. S. Bauer, Reo Bennett, Dick Bills, Maria Bolsey, P. D. Bond, Albert H. Blum, Dr. A. K. Chapman, L. W. Coulson, M. H. Donahoe, Jr., R. D. Frieke, W. S. Fujimura, Lois Gauch, M. L. Green, Mrs. Ellen Halleran, Donald P. Hochgreve, T. T. Holden, John Hood, Dr. Randolf Kingslake, Charles E. Kennard, F. B. Mehlenbacher, T. H. Miller, Vick Moise, Beaumont Newhall, Morgan Oats, Joseph Pignone, Joseph Price, D. C. Ryon, H. W. Sapp, Barbara Ann Smith, William L. Weeden, George C. West, Mike Wolk, W. A. Young, Karen Nelson, and A. L. Terlouw.

Hollywood, California
Rochester, New York
January, 1972

KALTON C. LAHUE
JOE A. BAILEY

Contents

Illustrations

Introduction

By the mid-1930's there was evidence that the Great Depression might not have been fatal to the American economy. Employment rose by over four million in 1934, business failures were sharply reduced, and the purchasing power of industrial workers rose by 25 per cent, aided in part by Ford's "5-dollar day." People were doing things again—48,000 of them watched Max Baer win the heavyweight championship from Primo Carnera. *It Happened One Night,* a screen comedy starring Clark Gable, was drawing the public into the theaters, and "Anything Goes" was at the top of the hit parade.

Businesses were beginning to take chances with new products which they hoped would brighten sales. A small maker of electric clocks in Chicago tried it with music and introduced the Hammond electric organ, while in Ann Arbor, Michigan, a group of businessmen risked their all to make the first table-model radio that could be plugged into either AC or the still-popular DC current; fortunately for our story, it sold rather well.

In 1935, as the economy continued to improve, some of the New Deal legislation designed to aid it came to grief. But "We Do Our Part" seemed more the spirit of the times than the slogan of the ill-fated National Recovery Act. Not all of 1935 was bright. Millions mourned the untimely death of the spokesman of the era, Will Rogers. Perhaps it was symbolic of the year that Alcoholics Anonymous was organized in New York City.

Then, in 1936, Franklin D. Roosevelt was elected to his second term, and there were signs that happier times were returning indeed. New York began planning a World's Fair for 1939. In the West, Hoover Dam was completed. Halfway across the world the German airship *Hindenburg* prepared for its first scheduled trans-Atlantic flight. For those who preferred more solid comfort, the H.M.S. *Queen Mary* embarked on her maiden voyage.

But most Americans stayed at home, reading *Gone with the Wind* and eagerly anticipating the movie version, when they would gasp as Gable spoke the screen's first "damn." A more lasting, more personal, and, for thousands, more exciting event was the appearance, almost by chance, from that small radio manufacturer in Ann Arbor, of the grown man's ultimate gadget. It was a small, easily manufactured, and inexpensive product, and it appeared that it just might sell—the first American 35mm camera.

With the dawn of 1937 the photographic industry realized that the depression was ending. The evidence was everywhere—in products, in interest, and, more important, in sales. Dealer organizations which had been rather loose-knit began to consolidate. The new fair-trade laws gave the small dealer what he hoped would be an end to price cutting, and as each major manufacturer added his support to the idea, a new wave of optimism was clearly reflected in the trade press. In retrospect, the fair-trade laws were more of a nuisance than a help, but in 1937 anything that gave hope was welcomed.

As an indication of the nation's relaxing of tensions, 1937 ushered in an age of the gadget, not only in photography but also in other industries. The best-known gadget supplier was Johnson Smith and Company, of Detroit, which offered its "huge catalog of 8,000 novelties" absolutely free (for a 3-cent stamp). To eager readers it was like a miniature Sears, Roebuck catalog, declaring, "We are in business for fun!" Its pages included such photographic wonders as a picture-in-a-minute camera ("Make money with your own business!") and a host of books and gadgets for the photographer. What matter that the books seemed to have been written at the turn of the century? After all, they cost only 5 or 10 cents. But then, for 10 cents you could also buy an electric-motor kit, and a quarter would bring a joy buzzer. Small cameras were very much in evidence—the ULCA, the UniveX, the Rolls, and, if you could afford to spend in the upper brackets, the Falcon Deluxe miniature with *f*/4.5 lens for $15.95. A kit containing a camera, film, an enlarger, and a complete darkroom setup was available. Kids speculated about what it would cost to buy one of everything in the catalog, while parents noted that the minimum order was 15 cents and silently thanked Johnson Smith and Company for at least frowning upon C.O.D. orders. They were busy with a new camera.

Part I The 35mm and Photography

1 The New Wave Arrives

Although not an American invention, photography gained mass appeal in the United States through simplification. From Kodak's roll film to today's automated instant-loading cameras, making photography serve the consumer has been a uniquely American attitude, copied but seldom initiated abroad. The usual procedure has been to take an idea and turn it into a practical product by innovating around those elements which seem to hinder more than they help. This approach was partly the result of competition from foreign companies, whose inexpensive labor forced American manufacturers to find simpler ways to do things more easily. Thus the Leica camera began as a complex, expensive German invention, while the first American 35mm was a simple, inexpensive design aimed at a depression-ridden consumer.

For millions of people the practice of photography had been liberated from the dark ages in 1888 by George Eastman, who assured them, "You press the button—we do the rest." Loaded with sufficient film to take 100 pictures, Eastman's No. 1 Kodak sold for $25 and was to be returned to the factory for film development, printing, and reloading at a cost of $10. The process gave rise to Kodak's claim to offering the first complete "system" of photography—and creating the amateur photographer. It was also possible for the owner to unload the film himself for shipment back to Rochester, or for home processing by the hardy few who wanted to try it. Each week Kodak received 300 to 375 cameras containing 30,000 to 37,500 negatives. Kodak estimated that more than 200,000 No. 1 Kodak cameras were manufactured during its design life.[1]

But it was the tinkering of an obscure German technician some three decades later that resulted in an entirely new group of photographers between the button pressers of the Kodak camera and the professionals. The Leica miniature camera designed by Oscar Barnack arrived in the mid-1920's and was responsible for spurring research and ultimately expanding one segment of the photographic market into a multimillion-dollar enterprise.

A small and remarkable precision instrument, Barnack's camera was an overnight sensation when it was introduced at the Leipzig Fair in 1925. As a result of this public acceptance the company of Dr. Ernst Leitz, in Wetzlar, Germany, prospered through the worst of the depression years. Sales doubled yearly despite a steady rise in prices and the restrictive import regulations which nearly strangled world markets and international trade.

Barnack had joined E. Leitz, Inc., in 1901, beginning in the microscope department. He dreamed of inventing something which would make him financially independent and in 1914 conceived the idea of a small camera which could be used to make test exposures for the movie industry. Motion pictures had adopted the 35mm format a decade before,

[1] It should be noted that, while it was called the No. 1 Kodak, it was not Eastman Kodak Company's first commercial camera.

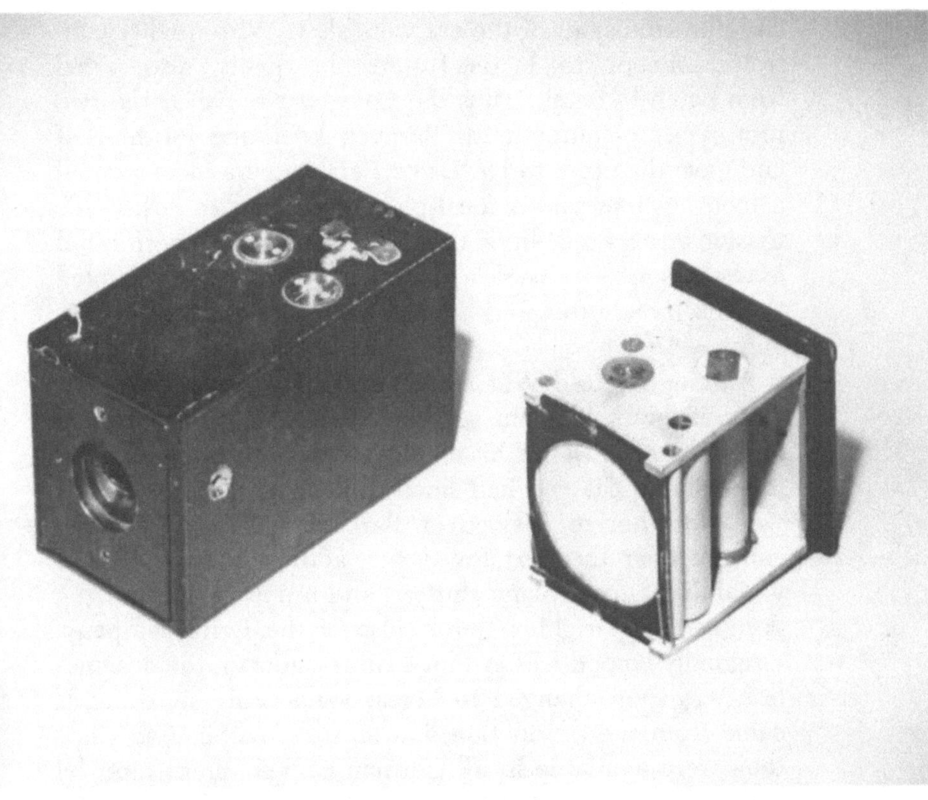

The No. 1 Kodak camera, often mistakenly identified as the first Kodak camera, was actually the first to use roll film in a manner practical for the amateur photographer. The camera placed photography within reach of everyone as a hobby because it removed the problems of plate handling, exposure setting, and film developing. An amateur could develop the film if he so desired, and the instruction book gave full instructions on how to do it. But most amateurs sent camera, film, and all to Rochester. The V on top of the camera was the viewfinder, although an accessory finder was made available later.

and film emulsions of the era were slow. With quality control a concept still in the future, film quality also varied from batch to batch. Thus the Leica was born. Of the two prototypes manufactured, Barnack kept one for himself and gave the other to Dr. Leitz. Each camera incorporated a focusing lens and a focal-plane shutter with adjustable tension and a fixed slit of 1⅝ inch. The use of the perforated movie film made possible a film transport which was coupled to cock the shutter and advance a frame-counting mechanism.

Although World War I interrupted Barnack's work, Dr. Leitz liked his camera so much that, once the war had ended, he encouraged his employee to continue developing it. Believing that he had finally hit upon the idea which would put him on easy street, Barnack perfected the basic concept over the next few years, adding a viewfinder, a variable-slit focal-plane shutter, and a new lens, the *f*/3.5 designed by Dr. Max Berek, also of the Leitz company. Originally named LECA (for LEitz CAmera), the designation was soon changed to Leica. While only six cameras came from the production line in 1924, within two years they were available in all German camera shops, and by 1928, when luxury items were riding a wave of popularity, Leica cameras were at the forefront.

But the idea of taking pictures with small cameras was not new with Barnack. It dated back to the 1850's in the United States. Often called "detective cameras," many cameras using small negatives were constructed in the forms of clocks, telescopes, watches, walking sticks, handbags, and the like. Concealment seemed to go hand in hand with the small-camera concept. One of the most popular was the canteen-shaped Stirn Vest Camera, designed to be strapped

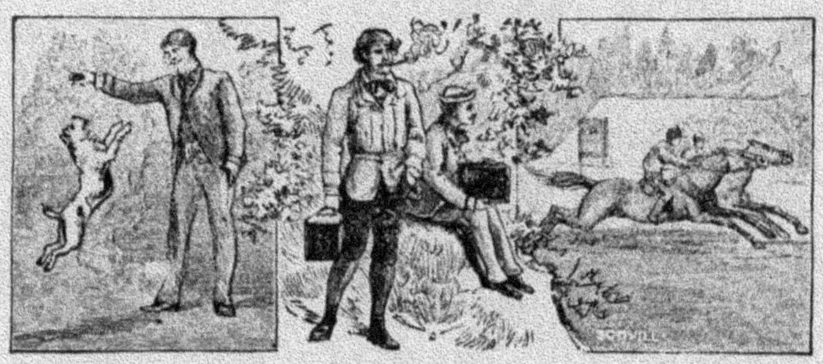

From the earliest days of photography there appears to have been an interest—perhaps generated by advertising copywriters —in taking pictures sub rosa. As this early Scovill ad shows, however, the cameras of the 1880's were hardly unobtrusive.

to the user's body underneath his waistcoat. Its lens peeked out through a buttonhole. It took six round pictures each about the size of a silver dollar. The Vest Camera was invented in 1885 by Robert D. Gray, an American lens maker. The following year Gray sold his camera to Carl Paul Stirn, who manufactured it in Berlin. While it was popular abroad, Stirn never introduced it commercially in the United States, though several appeared here. In 1905 the Victrix camera arrived from Germany, followed by many others during the next decade, each meeting with some degree of acceptance by the public.

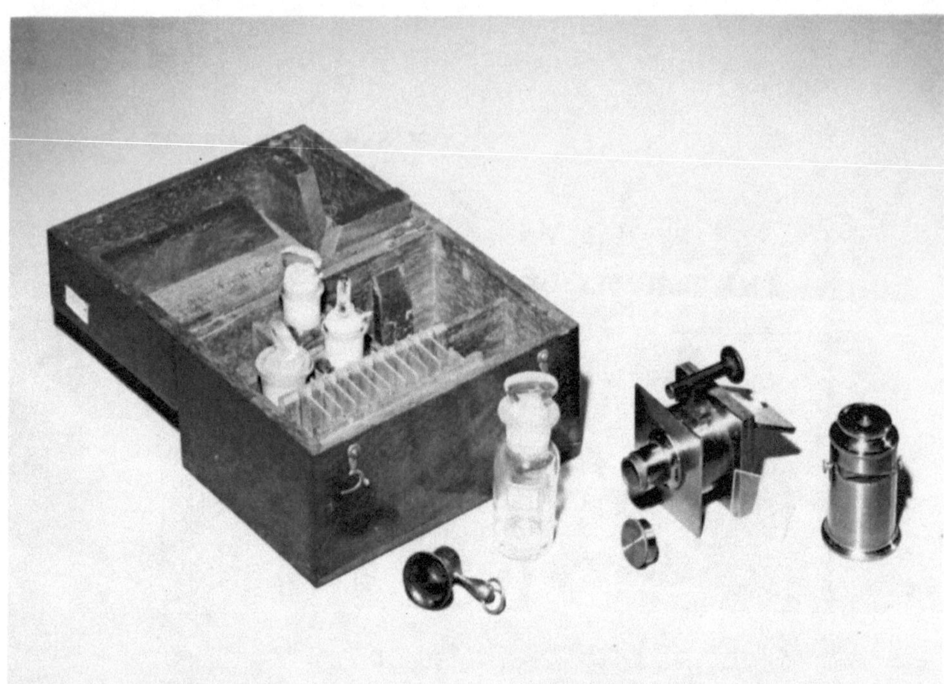

Not long after the birth of photography came a desire for small cameras. This Bertch Miniature Wet Plate outfit of around 1860 used 1-inch plates in a camera that was not much larger. While the idea was good, it is hard to imagine the usefulness of a "pocket camera" using wet plates.

While camera fans have generally accepted the Leica as the first still camera to use 35mm movie film, it actually was not.[2] At least two such cameras appeared on the American market as early as 1913–14: the Simplex Multi-Exposure and the Tourist Multiple. Little known then and apparently not remembered today, the Simplex Multi-Exposure camera was manufactured by a then-famous maker of 35mm motion-picture projection equipment. It had a remarkable list

[2] According to photographic historian Beaumont Newhall, the very first 35mm camera with the double-frame (24 × 36mm) format was handcrafted in 1912 by George P. Smith, of Richmond Heights, Missouri.

of features for its time and an amazing versatility that put even the Leica to shame.

Made of aluminum, steel, and brass and covered with leather and a black metal finish, the Simplex had an *f*/3.5 Zeiss Tessar I.C. lens and a OO ("double-oh") Compound shutter with speeds of 1 to 1/300 second, "time" and "bulb" (*T* and *B*). It used a reloadable 50-foot daylight-loading film magazine and could be adjusted to take 800 single or 400 double-frame pictures on one 50-foot length of film. There were at least three models of the Simplex, all of them similar in appearance and picture capacity. For a camera that offered so much and sold for only $65, it is amazing that its manufacturer saw such limited use for it. While the Simplex was advertised to appeal to "travelers, sportsmen,

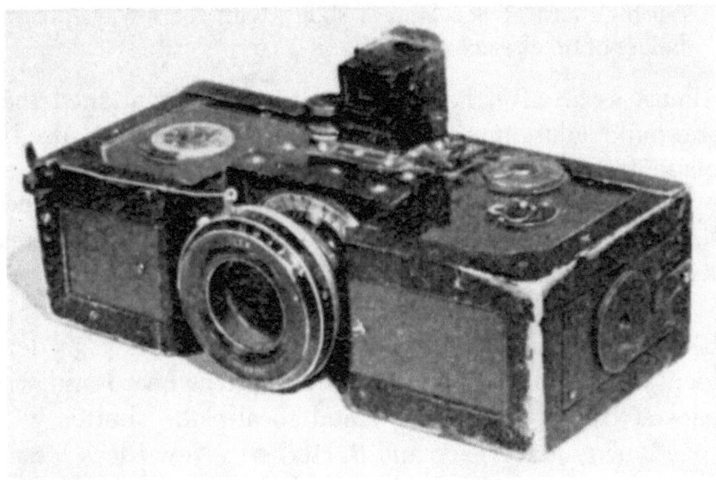

The 1912 hand-crafted 35mm of George P. Smith, said by photographic historians to be the first 35mm designed in the United States.

explorers, tourists, and amateur photographers," it was primarily exploited as a low-cost way of making lantern slides, which were then in vogue:

> The inventors of this camera have seen great possibilities in the use of standard perforated motion picture film (which is universally obtainable), keeping in mind the best ultimate use of the Multi-Exposure negative strips, viz.: the making of positive films from same and projecting these positives on the screen, step by step, similar to the use of slides in a stereopticon, with these very great advantages: Pictures are shown in the same consecutive order in which the original views were taken, being on a continuous band of film. A roll of fifty feet of film containing 800 views weighs but a few ounces, may be put in a coat pocket, is unbreakable, costs about six cents per foot ready for the projector, as against the bulk, weight, and high cost of 800 lantern slides, with the always-present danger of breakage.

Almost as an afterthought, the directions mentioned that "bromide enlargements from 4 × 5 to 11 × 14 may also be made from these small negatives with excellent results, while contact prints, especially from the double, or '400,' size, make attractive miniatures, just right for the latest stamp picture fad."

Henry Herbert's Tourist Multiple camera was another fascinating design years ahead of its time. Its *f*/3.5 lens focused to two feet on a rising-falling lens board and was backed by a self-capping metal focal-plane shutter with speeds of 1/40 to 1/200 and *B*. Herbert's New Ideas Manufacturing Company, of New York, produced the Tourist Multiple to use short ends of 35mm movie film in special magazines, giving 750 single-frame (¾ × 1 inch) exposures a load. The Tourist Multiple was doomed by the poor quality of available film and the advent of World War I, which closed down production for the duration and ruined the tourist market for which it had been designed. Of the 1,000 cameras sold at $125 each only a few remain, all in the hands of collectors.

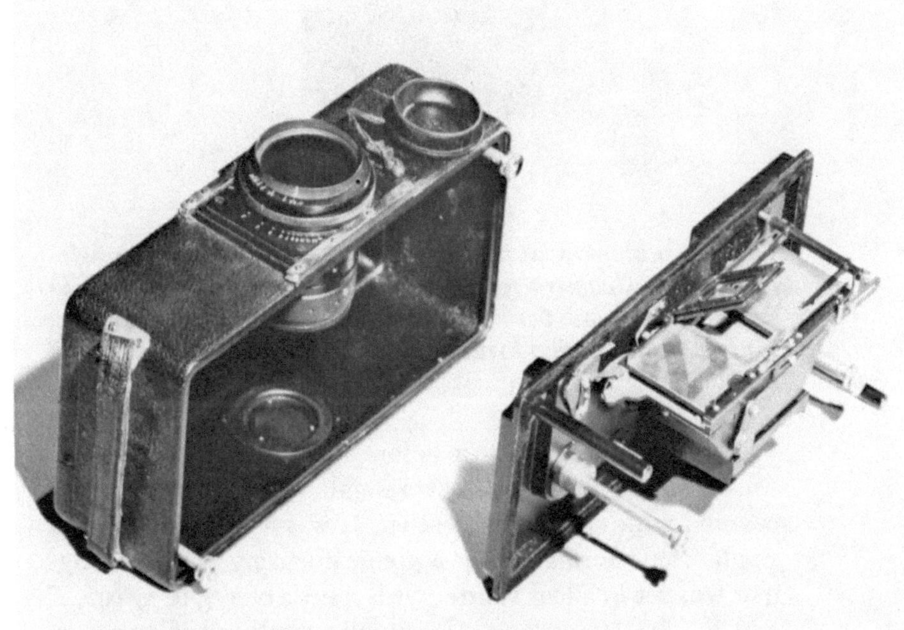

The Tourist Multiple followed the Simplex and is sometimes mistakenly said to be the first American 35mm camera. While well constructed, the camera was large and bulky, resembling a movie camera. The Simplex was much more versatile; the user could remove the exposed portion of the film without having to develop the entire roll.

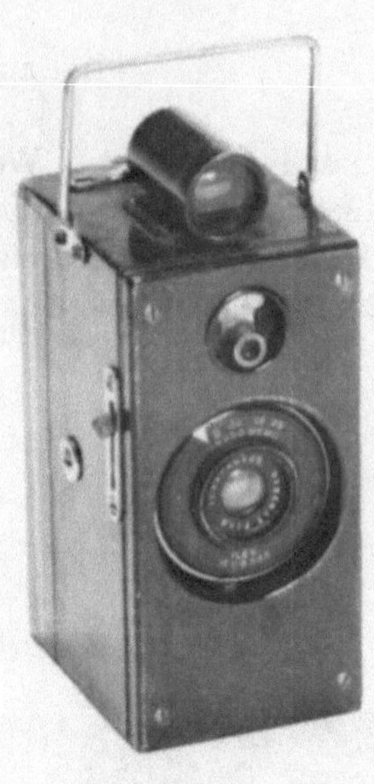

The Ansco Memo was introduced in 1924 as an upright single-frame design constructed of wood. At first it was finished in stained wood and varnish; later the wood was covered with leather. A unique film-loading system used a square box to hold the film.

Introduced in 1924 just before the Leica appeared, the Ansco Memo I camera was a small vertical wooden box covered with black grain leather. It was available only in a single-frame format using a 30mm focusing *f*/3.5 anastigmat lens set in a leaf shutter with speeds of 1/5 to 1/100, *T* and *B*. The Memo I was eventually marketed in two versions, with fixed-focus or focusing lens.

The Memo I incorporated several unique features: a cassette system similar to the European Rapid system which briefly challenged the Kodak instant-loading concept in the mid-1960's. It also had a claw-tooth-drag film transport with a spring-loaded return which pushed the film into place and

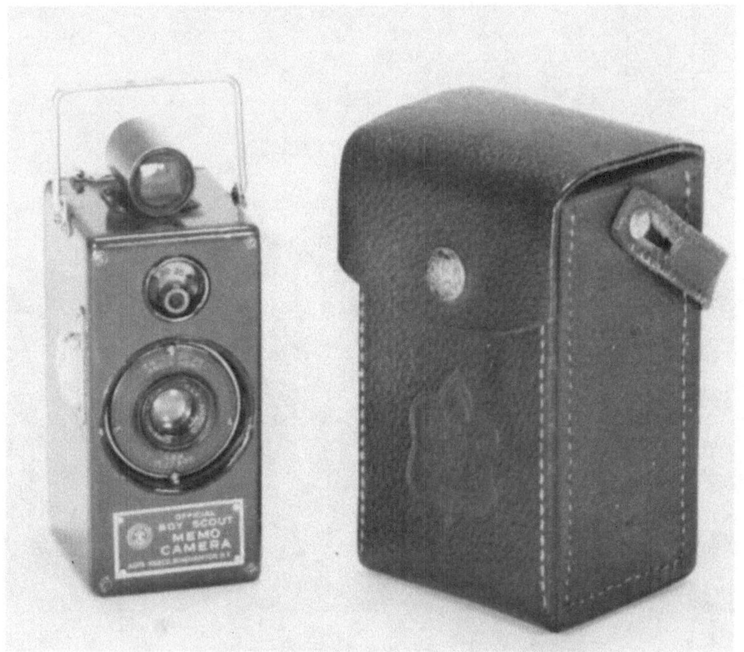

Like most other major small-camera brands, the Agfa Ansco Memo was produced in a Boy Scout model. Some were also made for Girl Scouts and Campfire Girls. This example is an f/6.3 model and is Boy Scout green in color, as is the leather case, which has slots so that it can be carried on a belt.

a focus-adjustment setscrew on the side. A full range of accessories, from Memo filters to a Memo projector, film printer, copier, and enlarger was available. Ansco published the monthly *Memo-Random*, to which registered owners were given a free six-month subscription. The original camera cost $20, including a soft suede-leather pouch. The fact that the Memo never enjoyed more than a modest degree of popularity is usually accounted for by its rather ungainly appearance and the not-greatly-improved quality of motion-picture film then available.

Why, then, did the expensive Leica camera catch the public fancy in a depression era? No one has a concrete

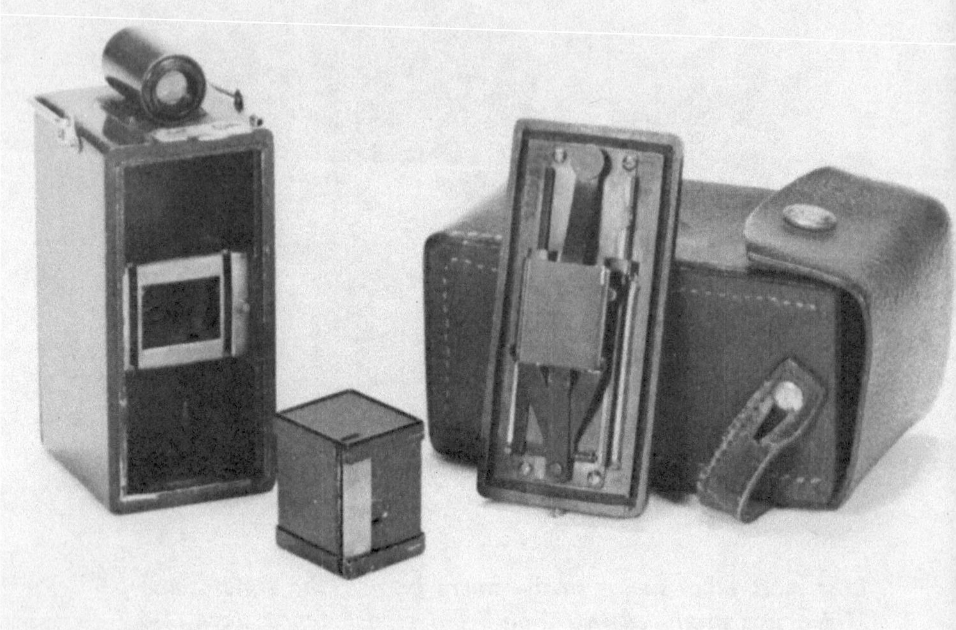

The Ansco Memo I introduced its own 35mm film-handling system, different from the Leica spool of the same era. While original models used small wooden boxes with metal ends, the wood box was later changed to all-metal, as this example shows. A small spring in the box tightened the film as it was fed into the chamber. The advantage of the Memo system was that it required no rewinding, and a single-stroke lever on the camera back brought new film into position. The Memo system was refined in 1936 to fit a rounded-corner metal camera and later became tear-shaped when it became the Agfa Karat system. In the sixties it was further refined into the German Rapid system. The importance of this development is impressive when coupled with the world-standard Kodak 135 magazine and the 126 cartridge. All major film-handling systems used in 35mm cameras (including the 828 roll film) evolved from American ideas and designs.

answer to this question, but one reason for the Leica's popularity is clear. Those who are completely immune to gadgets (especially new and attractive ones) are rare indeed, and Barnack's Leica was a gadget par excellence. People were proud to display the sleek-looking camera, with its many mysterious knobs and handsome trappings, and its extrinsic value quickly made it a vogue among the smart set. Wearing one around the neck soon became the thing to do, especially by people with money. Contrary to folklore, there was money in the depression years, and those who had it liked to show off their status to the world. The Leica did not resemble the traditional camera. It had no bellows, no bulky exterior, no focusing hood. It was small, was aesthetically satisfying in appearance, and could be slipped into a coat pocket for carrying. Mobility was a large factor in its success, and the Leica owner scorned posed photographs—he shot pictures when his subjects were unaware of the camera's presence, and the new school of candid photography came into being.

In public acceptance the Leica had no competition until the Rolleiflex camera appeared in 1929. This twin-lens reflex used 120 roll film (originally 117), producing 12 pictures measuring 2¼ × 2¼ inches, and thus did not compete directly with the Leica in use or appeal.[3]

The toughest competition for Barnack's camera came from the Zeiss-Ikon Contax camera, introduced by Carl Zeiss, Inc., in 1931, and even that camera faced an uphill struggle against the entrenched Leica. By 1936 only 40,000 Contax cameras had been sold in comparison with sales of 210,000 Leica cameras. Of these, Contax sold 10,000 and

[3] By mid-1936, 120,000 Rollei cameras had been manufactured and sold, about half the number of Leicas.

Leica 25,000 in the United States, which was not then particularly noted as a camera-conscious nation. Nevertheless, while most foreign sales were made in Japan, Great Britain, France, and South America, the miniature-camera concept was rapidly taken up by Americans. Librarians and researchers used 35mm cameras to copy old manuscripts and documents, doctors and dentists recorded their patients' ailments on film, scientists clamped their 35mm cameras to microscopes, and travelers brought back increasing numbers of pictures. Many of these photos were published by newspapers and magazines, and the press began to equip its staff photographers with the new miniatures. The *Milwaukee Journal* was one of the first to do so, and while Rollei cameras (made by Franke & Heidecke) and Speed Graphics (made by Graflex) remained the primary tools of the *Journal*'s photographic reporting, its use of the 35mm gave the camera an air of respectability and professionalism early in the game.

Neither the Leica nor the Contax was inexpensive to buy, own, or operate. Although a Leica could be bought for as little as $98, the typical sale rang up close to $200, which included the *f*/3.5 Elmar lens, a particularly desirable one because of its reputation for sharpness. The Contax with a comparable lens sold for $245. A decade after the Leica's introduction to the market, by Leitz's count 415 accessories were available for it. To some extent every camera purchaser was soon the captive of the gadget craze which accompanied the 35mm boom.

In 1932, with eight years of momentum and a well-planned advertising campaign behind it, the Leica moved into the American market. The story might well have ended there had it not been for the acceptance and development

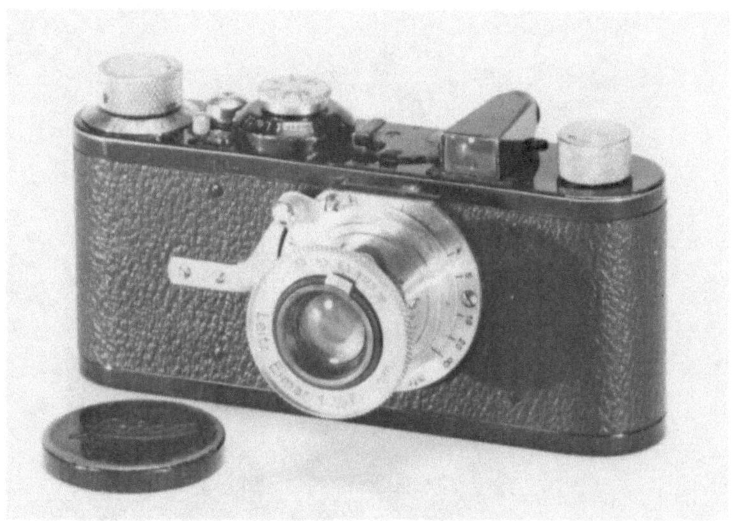

Why is the Leica I (above) considered the camera that launched the 35mm era, when the Ansco Memo, introduced the same year (1924), is hardly remembered? There may be many answers: the quality and the appeal of German technology, or perhaps the unique focal-plane shutter and the quality of the lens. On a price basis, the Memo was a much better buy at about one-tenth the price of the Leica, and in total sales the Memo outranked its rival. Yet today the Leica is universally recognized as the father of modern 35mm cameras, while the home of the Memo (Ansco) no longer even manufactures cameras, and the United States no longer makes 35mm cameras.

of the 35mm in America, for prices of German cameras crept up 30 per cent after 1932 to compensate for the devalued American dollar, and in July, 1935, an additional 25 per cent was added to offset a United States tariff increase. The German government subsidized its camera-export in-

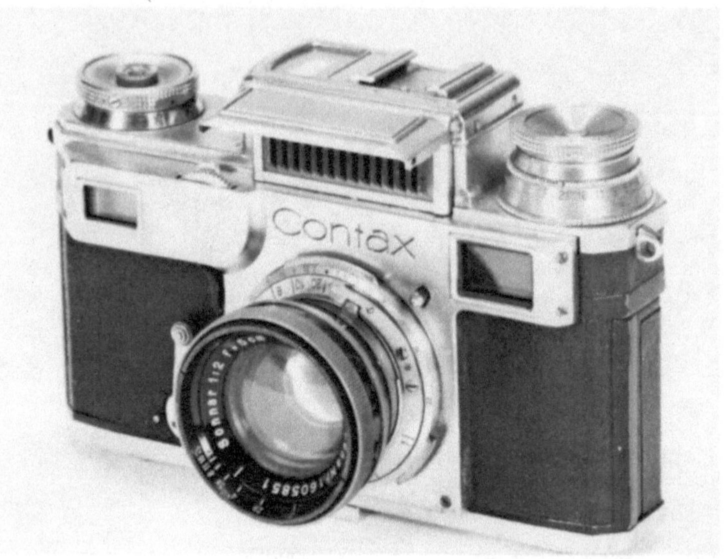

The Contax was the other quality camera that attained success in depression-ridden America. The model IIA was exactly like the IIIA (shown), except for the attached meter. Meter technology left much to be desired, and almost all IIIA's had problems. Rumors that the vertical metal shutter was also problem-prone hurt sales of the camera. In fact, the cotton ribbons did occasionally break, but not with the reported frequency. The camera had the virtue of working well in cold weather, which the Leica did not, and at the start of World War II the Contax outsold Leica for a short time. After the war the Zeiss plants in the Russian Zone were dismantled and shipped to Eastern Europe. A West German Zeiss produced the IIA again, but it never attained the success of the Leica—whose plants had been left virtually intact despite Allied air raids. In August, 1971, Zeiss announced that it was abandoning production of amateur cameras. Except for the Leica, this announcement marked the end of 35mm cameras from Germany.

dustry after the dollar devaluation. Although the tariff was removed one month later, when Germany agreed to drop its subsidy, German camera manufacturers did not drop their prices to the former levels; the price increase was retained to compensate for the loss of the subsidy. In the face of these developments some trade journals expressed fears that the 35mm-camera business was doomed, a prophecy which industry soothsayers had been uttering from the outset. Many felt that the general public simply would not accept as more than a passing fad the small-size negative with its inherent limitations. As it turned out, the soothsayers were wrong.

2 A Passing Fancy?

Readers who do not recall the 1930's will find it difficult to understand the revolutionary effect that the 35mm miniature camera had on photography. One really must have lived through the period to understand the impact this instrument had on photography as a hobby during those lean years. Strangely enough, few people in a position to capitalize on the vast interest created by the minicam believed that it was more than just a passing fancy, akin to the periodic interest in stereo photography and the ultra-miniature craze of the early 1950's. Established professional photographers considered that negatives smaller than 5 × 7 inches were for amateurs and held staunchly to their belief that the 35mm bubble would burst quickly, leaving many owners with useless cameras and equipment which would only gather dust. They regarded the 35mm somewhat as many of today's photographers regard ultraminiature size. As a result of this general attitude there were no outstanding guiding geniuses of the 35mm movement.

A highly respected member of the photographic fraternity, Dr. Kenneth Mees, of Eastman Kodak Company's Research Laboratory, would have been the ideal person to lead the bandwagon. Dr. Mees was brought to Kodak in

1912 when it purchased the English firm of Wratten and Wainright, manufacturers of sensitized products and filters. For many years he firmly believed that what the public really needed to inflame its photographic interest into a roaring passion was an "Aunt Molly" camera, one which even a fumble-fingered amateur could operate and enjoy. (While at its introduction the 35mm camera did not completely meet Dr. Mees's requirements, the initial reception and later refinements would ultimately contribute to the appearance of such a camera.) The eminent researcher was wrong in his refusal to take the movement seriously. A purist of the old school, Dr. Mees stood firm against the 35mm format, mainly because of what he felt to be the inherent inadequacies of the small negative size, referred to with some disdain by large-camera adherents as "postage-stamp negatives."

But postage-stamp negatives and the cameras which produced them had a strong case for public acceptance, and once the smaller manufacturers, who in this case snatched the lead from the giants of the industry, discovered that they did indeed have the public's attention and interest, they made the most of the advantages of 35mm in their advertising. While no mammoth campaigns such as we are accustomed to today swept the country, photographic advertising did get around in the 1930's; word of mouth was as important as any ad campaign could have been. When your next-door neighbor came home with a new 35mm camera slung around his neck and a few days later invited you to see the results of his weekend with the camera, a peculiar taste crept into your mouth, and your eyes narrowed to a fine slit, focusing upon the bright glint of the glass, brass, and chrome he so lovingly fondled before you.

For a segment of the American population in the depression years new cameras excited almost as much interest as new automobiles do today. There was a certain status in possessing one and certainly in repeating the incident which had converted you—that is, inviting your next-door neighbor over to look at your pictures. A large degree of ego satisfaction was also involved.

In those years photography was not the widespread hobby it is today. Professional photographers remained more or less mysterious to the general public, and few people knew how they worked or what went on under the dim red light in the smelly room behind the curtain separating the studio from the "laboratory." Amateur photographers, if

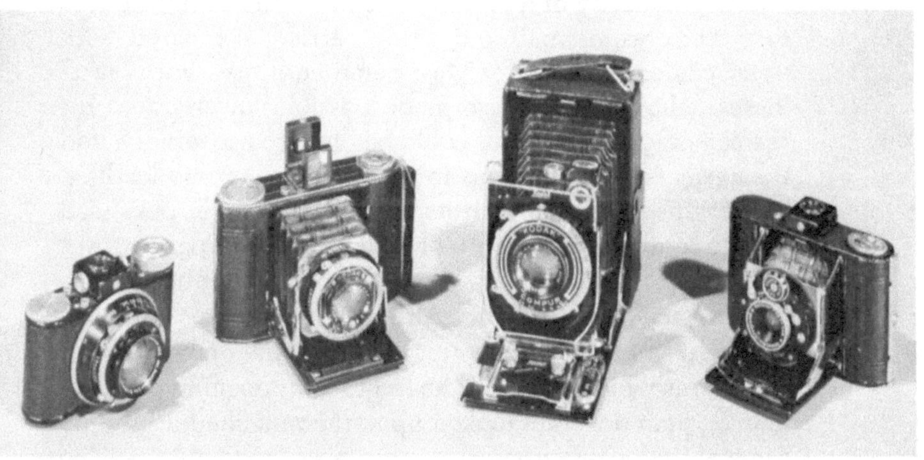

What kinds of cameras were called "miniatures" in the late 1920's and early 1930's? The cameras in this line-up (produced by Kodak's Nagel works, its quality-camera plant in Stuttgart, Germany) were all considered miniature cameras. Left to right: the 127 Pupelle, the Duo Six-20, the Recomar, and the Vollenda.

one could call them that, were for the most part a Sunday snap-shooting group, who took Dad's Autographic Kodak camera or their own Kodak Brownie box camera on a picnic, made pictures filled with distant groups of relatives (which today look like police lineups), and left their film with the local studio to "soup" and "print" during the week. That is not to say that there were no amateur photographers who did their own darkroom work. Amateur darkroom equipment and supplies were a substantial part of Kodak's business even then, and Kodak was only one of dozens of suppliers. But the acceptance of the 35mm camera expanded this group of amateur photographers into a sizable number of camera users.

The popularity of 35mm cameras was due to several factors. They were small and easily carried, compared with other cameras available. Most came equipped with "fast" lenses, whose smaller size made possible lower prices for the cameras. The cameras could be put into use rapidly, and their users could make up to 36 exposures in one loading. The high-speed lens, with its short focal length, possessed an inherently greater depth of field, giving sharper pictures than their larger brothers even with less attention to extremely accurate focusing. Although their users soon disregarded this asset by shooting pictures under adverse light conditions which required the largest lens opening possible, many 35mm devotees looked upon the rangefinder used by some of their contemporaries as a concession to old age. The inability to judge distance correctly was considered the mark of the snapshooter, whose Brownie or UniveX could not care less one way or the other. Manufacturers took advantage of this built-in snob appeal for many years, advertising that the extreme depth of field of the 50mm lens allowed satisfactory focusing without the use of a rangefinder, a claim which many experts seriously questioned.[1]

In the mid-1930's the 35mm miniature camera was responsible for the formation of about 60 camera clubs spanning the United States from Boston to San Diego. While clubs were not completely dominated by the 35mm devotees in the beginning, nonetheless many of them came to life bearing the name Leica. The first club was organized in the United States in San Diego in 1931. Five years later the 6 largest clubs in the country boasted the following active memberships: Detroit, 400; New York, 157; Philadelphia, 150; Boston, 87; San Francisco, 75; and Chicago, 50. The average membership of the remaining 50-odd clubs was estimated at 40. The membership dues, which ranged between $4 and $15 annually, meant that membership kept fluctuating in a depression society. These clubs sponsored religiously attended programs, which included courses in composition, pictorial photography, lighting, and developing techniques; exhibitions of members' pictures; guest speakers; and weekend or holiday outings.

New York numbered several miniature-camera clubs in its several boroughs, and weekend tourists were all but overwhelmed by excursion boats and buses searching out new subjects for the omnipresent 35mm cameras and their owners. No part of the city was safe from outings organized to stalk new angles of the local panorama.

Some clubs served as departure points for a few members

[1] Although 50mm lenses are standard, theoretically, the more nearly correct lens-to-film size is 44mm. Leitz established the tradition of the 50mm, and other cameras offered with shorter lenses are sometimes referred to as "moderate wide-angle." The public readily accepted the 50mm, and when Kodak brought out the first Signet with a 44mm lens, users wrote letters of complaint about the "too-short lens" (see Chapter 3).

who formed their own elite organizations. The best remembered was New York's Circle of Confusion Club, which met on the second Monday night of each month and rapidly became the nation's number-one minicam group. Since the club counted four professional photographers and eight photographic distributors among its members, the competition within the organization was extremely serious, and members won photographic contests and held salon exhibitions far out of proportion to the length of the membership roll.

There were also unorganized 35mm fans, who far outnumbered their club counterparts. Many yearned to be professional photographers and did not want to waste valuable time or money affiliating with organizations when they could be out searching for pictures which might open the way into a new occupation.

Photographic manufacturers went to great lengths to spur to greater glory those customers who bought their cameras, chemicals, or films. Picture competitions were constantly being offered for photos taken with a Leica, a Contax, or an Argus camera or for those captured on a particular brand of film, perhaps even with the aid of one of the new flash synchronizers, which appeared on the market in the middle 1930's. Manufacturers of other kinds of merchandise also entered the spirited game. The Royal Typewriter Company was especially fond of pictures showing its new portables and offered several competitions for photographs suitable for advertising use. Prizes, which often exceeded $250, were offered at regular intervals for winning photographs, providing a strong incentive for the taxi driver or sales clerk whose weekly salary was stretched to its limit to cover the cost of living and still leave enough for a few rolls of film and some printing paper. For the fortunate or talented, winning a prize of that magnitude often equaled a half year's salary in his regular job (if he had one) and was a great deal more fun than work. And there was always a chance that a prize-winning picture might open up a new career for its maker. The annual Leica exhibitions in New York provided recognition for Leica owners, who vied for invitations to

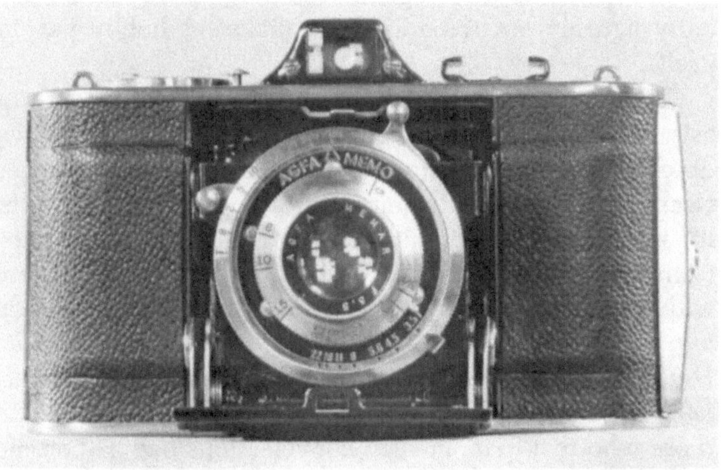

Under Agfa management, Ansco developed a more modern approach to 35mm design in 1935 with the metal Memo models. A revised film cartridge, destined to be the Rapid system of later years, was introduced with this camera, but unfortunately 1935 was also the year in which Kodak introduced Kodachrome color film, and it was available only in a different style of cartridge. The newly designed metal cameras used a more traditional horizontal format and were available with an f/4.5 or an f/3.5 lens in single- or double-frame size. Agfa Ansco never produced a "standard" 135 cartridge camera in this country, although postwar cameras were imported and sold under the Ansco name.

hang examples of their skill. Such exhibits proved to be popular with the public; the 1935 show in Rockefeller Center lasted two weeks and drew more than 25,000 spectators.

Another source of income was derived from taking pictures of drug- or department-store displays featuring specific products. Most pictures of this nature, if acceptable, were worth $5 from a manufacturer, or at least a sample of his product. The truly ingenious photographer had no difficulty figuring out some angle to make his hobby pay for itself.

The 35mm had other impacts on the lives of its enthusiasts. It brought forth an entirely new jargon, which was quickly incorporated into everyday language. When owners swapped stories about their latest attempts to immortalize life and nature, the uninitiated who listened in were likely to hear dozens of unfamiliar terms casually tossed around with knowing glances: *reticulation*, *gamma*, *spotting*, *toning*, *superpan*, *fogging*, *agitators*, *Newton rings*, *density*, *D-76*, *resolving power*, *emulsion speed*, *Scheiner*, *bromide*, *parallax*, *Champlin 16*, *latitude*, and many others. Few of those who took part in such conversations had a thorough acquaintance with the technical ramifications of all the terms they used, but it was an era of self-education, and many developed into competent photographers without benefit of formal training.

Publishers were quick to note this rapidly growing market and met the challenge. In a few short years they brought into existence more than two dozen new journals and publications for miniature-camera fans. These publications joined older magazines, of which *American Photography*, *Camera Craft*, and *The Camera* remained the most popular. All three had gracefully incorporated the 35mm format into their pages, and their vitality was amazing; the photographic magazines of the mid-1930's were virtually a continuing course in the fundamentals, techniques, and aesthetics of photography, with a liberal sprinkling of do-it-yourself articles.

But in the mid-1930's photography magazines began to change in emphasis and format. A new wave in typography and magazine makeup resulted in *Minicam*, *U.S. Camera*, and especially *Popular Photography*, which gave the reader a much greater variety of topics, more pictures, and, even more interesting, hundreds of advertisements of new and used products. Of all the photographic magazines which this country has produced, *Popular Photography* has survived and prospered as no other, becoming, in some people's opinion, the leading photographic journal not only in the United States but in the world. Many others surpassed it editorially and authoritatively, and most surpassed it in excellence of format. But no other as successfully combined an understanding of what the amateur wanted and how to satisfy his wants. *Popular Photography* introduced color illustrations and, for a camera-happy public, was the first to print a directory of everything photographic on the market.

Of all the American photographic magazines *The Camera* seems to have lived the longest. Begun before the turn of the century as *Camera and Darkroom*, it breathed its last in the early 1950's, when it was absorbed into *Popular Photography.* Throughout its many years the magazine reflected the progress of photography but remained to the last devoted primarily to the craftsman, a not-too-salable product in the end. England has probably produced the largest number of photographic journals and the most comprehensive ones. The English not only write about photography, they

CANDID CAMERA CORP. of AMERICA
ANNOUNCES THE NEW

Perfex FORTY FOUR

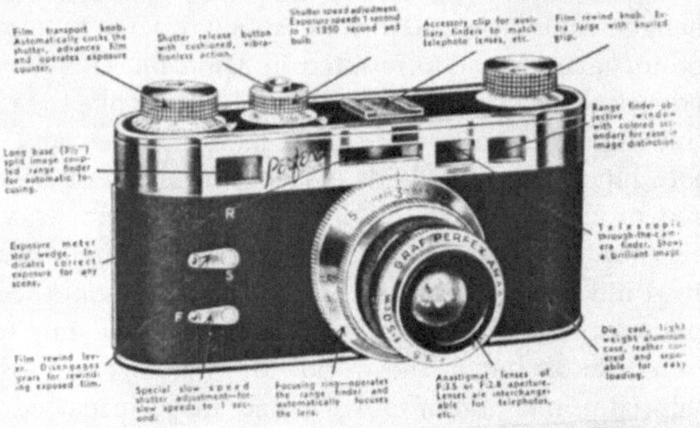

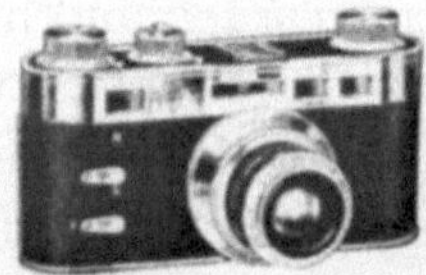

SPECIFICATIONS

Leather covered die cast aluminum case with chrome trim. Camera body measures 5 9/16 x 1 9/16 x 2 23/32 inches, weighs approximately 22 ounces. Loads with standard 35mm film, 18 or 36 exposure, for pictures in black and white or color. All parts are accurately machined and fitted, assuring smooth trouble free operation.

THE CAMERA OF YOUR DREAMS

The Perfex Forty-four, 35mm Miniature Camera offers you every proven and acceptable feature of the most expensive miniatures, yet, sells at a fraction of their cost. Study the features described above—the engineering accomplishments of coupled range finder, focal plane 1 to 1/1250th second shutter, built-in exposure meter, automatic film transport, *built-in synchronizer for flash light photography* and the convenience of 35mm film, 36 exposures per load. By any standard of comparison this American made Perfex Forty-four is an outstanding value. See it at your dealer.

WITH F:3:5 LENS
$37.50

WITH F:2.8 LENS
$47.50

CASE EXTRA $5.00

CANDID CAMERA CORPORATION of AMERICA
844 W. Adams Street Chicago, Illinois

This 1939 trade advertisement for the Perfex Forty Four also appeared in consumer magazines and illustrates the emphasis on technical information to sell cameras at the time. Today's ads lean heavily on what can be done with a camera, but during the heyday of the 35mm prospective owners were more interested in the features the camera offered.

dissect it. They have never quite reached the point of loving the equipment for itself, as do Americans, which may help explain why a home-photo industry is almost nonexistent in the British Isles today. But it was the English who first began a publication called *Miniature Photography*, and they did so before 1900.

Today England publishes the most technically oriented magazines, while the Germans and Swiss share honors for the most beautiful publications. But the world's largest circulation still belongs to *Popular Photography*, one of the three publications of the thirties which have survived the passage of time and changing public tastes. As for the other two, *Minicam* is now *Modern Photography*, and *U.S. Camera* became *Travel and Camera* and was purchased in 1969 by American Express, after which it all but ceased its photographic emphasis of the past.

Willard Morgan was one of the first writers to embrace the 35mm concept. He became a confirmed Leica camera user as early as 1928 and employed it to illustrate his travel and trade articles with candid pictures. In 1935, with the assistance of Henry J. Lester, he produced the first edition of the most durable and authoritative work on 35mm photography ever published, the *Leica Manual*. After countless revisions and subsequent printings, it still stands as one of the best books available on the subject.

Indicative of the importance of the 35mm was the recognition that camera owners were prime prospects for theater tickets. In the thirties, while the movies tried bank nights and bingo, stage-theater managers designated certain nights for candid shooting and filled their houses with camera fans. Actually, the managers were simply formalizing, regulating, and profiting from a practice which had started earlier in the decade. Minicam fans had been taking pictures, with or without the management's permission, for some years when the *Camera* arranged for a photographer's night at New York's Winter Garden Theatre in July, 1935. That night, much to the amazement of the management, about 800 avid photography fans, accompanied by more than 600 friends and relatives, descended upon the theater. They overran the performance of Earl Carroll's *Sketch Book Review*, stalking up and down the aisles, climbing onto the stage, and standing in their seats to find the proper angles. For this privilege each of them had paid between $2.20 and $3.85 admission on a hot summer night.

In September, not to be outdone, Radio City Music Hall opened its doors on the rehearsals of *Jumbo*, and for four weeks before the premiere minicam fans were allowed to prowl on and behind the stage during rehearsals, shooting to their hearts' content—for a small fee, of course. Soon radio broadcasting studios had also taken up the idea. The culmination of the fad came when George White took advertising space in photography magazines inviting readers to attend his 1936 *Scandals* at the New Amsterdam Theatre and shoot pictures of whatever caught their fancy. Managers urged those who wished to capture the shows on film to attend a regular nonphotographic performance on an earlier evening and "plan" their shots, a reasonable suggestion which also meant an extra admission at the box office. The only restriction placed on the photographers was that they should not take flash pictures, either with powder or with the new flashbulbs, and this restriction made the minicam king.

3 The Glass, the Brass, and the Chrome

By the autumn of 1936 the 35mm camera had taken a firm hold on the American public. Willoughby's, on Thirty-second Street in New York City, the largest retail camera outlet in the nation, reported to *Fortune* magazine that its camera-sales pattern showed the 35mm outselling every other type of camera available by a nine-to-one ratio. Of course, this particular sales pattern did not hold true in all photo outlets, but it was a valid indicator of the 35mm's appeal. In 1936 the profit potential was excellent for stores handling the Leica and Contax cameras. The manufacturing cost of a $375 camera was a mere $70, and the factory markup, or profit on its sale to the importer, was $47. A $64 duty and an excise tax of $25 made the importer's cost $206, to which he added his own $44 markup. Thus the dealer's cost was $250, and his profit on a sale amounted to $125, a difference which encouraged a healthy trade-in allowance and the development thereby of an important used-camera market. Some operators did even better. In the mid-thirties, a pawnshop owner near Madison Square Garden could get you any Leica or Contax camera you wanted at 50 per cent off list price. You simply went to a retail store and picked out the model you liked; the pawnshop owner would have one stolen for you.

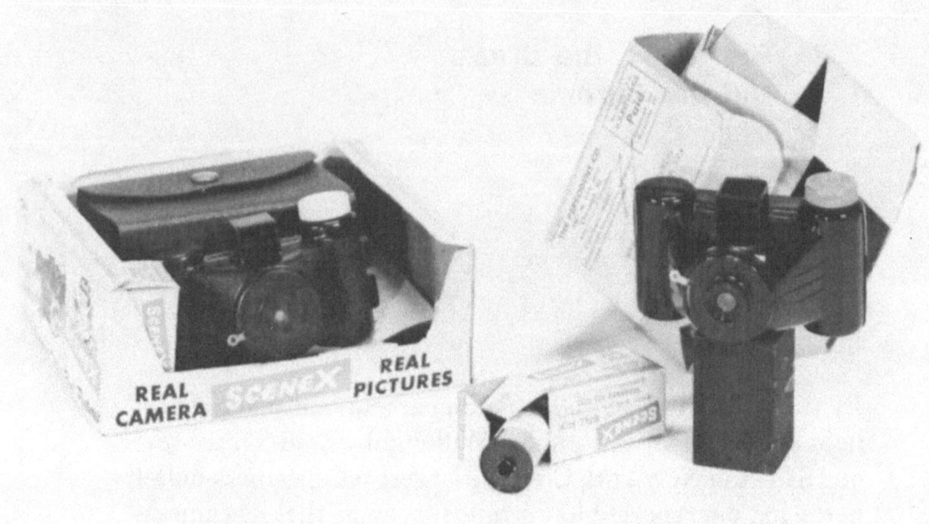

How many cameras were available for 15 cents—even with a top from a toothpaste box? Only the Cub (later sold as the SceneX for $2). A roll of 828 film cost more than the camera, although some stores sold it for 10 cents as a loss leader to gain the finishing business. The camera took acceptable pictures for a while, but unstable plastic construction soon took its toll: the backs warped, and light entered the camera unless it was taped together. Even so, for 15 cents it was a bargain hard to top.

In addition to the 35,000 Leica and Contax cameras which had been sold in the United States by 1936, there were about 45,000 35mm cameras of American manufacture in the hands of photographers. If we consider each of these cameras as representing a $14.00 sale (38,000 were Argus cameras, manufactured by International Research Corporation, selling for either $12.50 or $15.00 each, and most of the rest were Kodak Retina cameras retailing at $57.50 each) and each of the Leica and Contax cameras as costing $200, the grand total comes to just a shade under $8 million over an eight-year period, with the bulk of the sales coming in 1934–36. That was a tremendous amount of money for a nation which was just beginning to find its way out of the economic wilderness, and it was only a portion of the total amount spent for cameras.

Add to that figure the undetermined cost of accessories, film, paper, and darkroom equipment sold to the 90,000 devotees of 35mm, and the total becomes a staggering amount, even by today's standards. We can lightly pass it off with the old saying, "Where there's a will, there's a way," but members of the younger generation who do not recall the depression years of their fathers may well be justified in questioning whether the supposedly rampant misery which abounds in our folklore of the thirties was really as prevalent and widespread as some social historians have contended. We cannot answer that one, but it seems apparent that photography occupied the minds and emptied the pocketbooks of a great many people 35 years ago, perhaps even to the extent that they sacrificed other niceties of life for the whirring click of a camera shutter.

One of the 35mm camera's greatest appeals lay in the number of pictures which could be taken on one spool of film. Film for other popular cameras of the day allowed only eight pictures a roll, and for them a Sunday outing meant either carrying additional rolls of film (which were often overlooked or forgotten) or accepting the limitations of the camera. Though manufacturers had paid insufficient attention over the years to the problem of providing easy access to the compartments which held the rolls of film, such cameras were not especially difficult to load, unload, and reload.

But fumbling fingers were likely to botch the task and fog the edges of the film hidden beneath the paper backing. As usual, the members of the fair sex were the smarter. They restricted their photographic activities to using up the exposures on the roll in the camera and let their dealers unload and reload their cameras when they brought in the film for processing. For many years the 35mm cameras belonged to the men; women preferred the simple box or folding cameras—innumerable industry consumer surveys over the years indicated that women considered the 35mm too gadgety and complicated-looking. But according to a recent survey, the women photographers have once again gained dominance; at this writing industry executives attribute 53 per cent of the pictures taken by American consumers to the distaff side.

During the thirties and forties so many 35mm cameras were advertised with a select group of key phrases—"precision miniature," "perfect instrument," "designed by experts," "built with precision," "extraordinary quality"—that the phrases lost significance long before the period ended. But those years of glory for the American 35mm cameras were characterized by a freshness and vitality of design unfettered by preconceived ideas. It was an era in which the best attributes of the Leica and Contax, two cameras which lived up to the advertising wordage, could undergo a metamorphosis for the general public, winning approval in the sales of Argus, UniveX, Kodak, Clarus, Perfex, Zephyr, and many other cameras. Yet it was also the age in which the Kodak Ektra was conceived, born, and allowed to die and the Kardon and Foton cameras were rejected at the counter.

While some camera designers and industry executives

may well disagree, we believe that the golden age of the 35mm camera represents one of photography's most innovative periods. It was the age when the photoelectric meter was developed and coupled to the camera. Flashbulbs appeared, and flashguns were synchronized to shutters. Interchangeable lenses and backs, self-timers, rapid film winders,

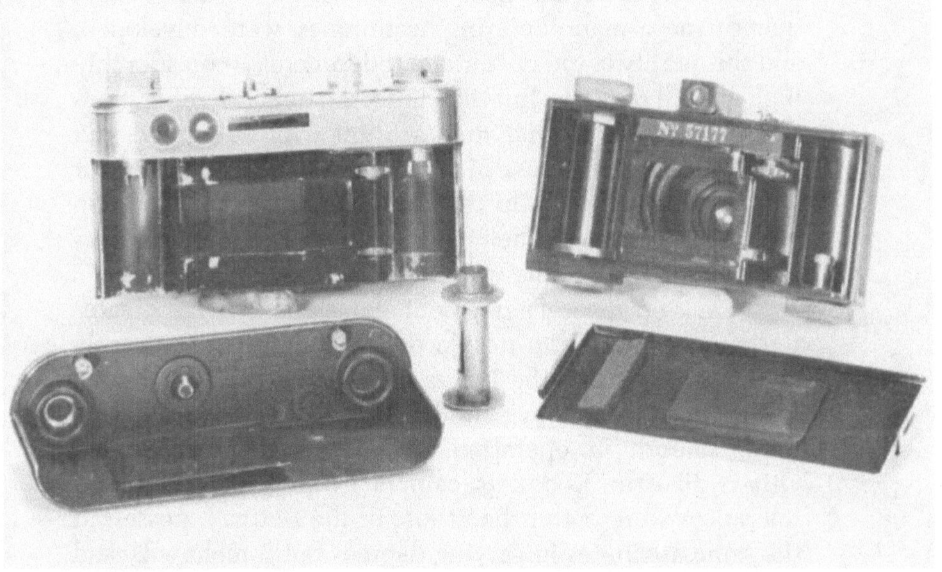

Two approaches to American camera design and manufacture in the thirties can be clearly seen inside the Perfex Forty Four (left) and the Argus Model A (right). With its focal-plane shutter, rangefinder, and meter the die-cast Perfex was heavy and somewhat crude. By contrast, the Argus was constructed of plastic and aluminum, with just a bit of brass in essential parts (note the single sprocket wheel, a sign of an early model). With its removable take-up spool, the Perfex was capable of magazine-to-magazine film loading.

coupled rangefinders, variable viewfinders, dual-focus lenses, and a host of other features, some which are proclaimed as new today, were available. The fifties were a decade of experimentation, followed by assembly of the worthwhile elements into an integrated whole in the early sixties and a repackaging of older concepts in more sophisticated designs during the latter years of the decade. Subsequently new manufacturing techniques were developed, and the quality of today's automated cameras is considerably higher in all respects, but they unfortunately resemble modern automobiles in that they are uniformly designed and have a sterile sameness of appearance and utility. All of which brings us back to the cameras under consideration and a discussion of their quality, whose resolution is as elusive as the will-o'-the-wisp.

There is no general agreement about exactly what creates the aura of quality in design or manufacturing, but early 35mm cameras like the Leica, the Contax, and the Kodak Retina possessed it in abundance. Each was sturdy, well built, smooth in operation, and long-lasting in design. Others, like the Kodak 35 camera (whose manufacturing tolerances were better than those of the Retina), possessed the same attributes in varying degrees but somehow lacked that mystical aura. But most 35mm cameras produced in the late thirties seem to have been designed with a deliberately high degree of "flash appeal," an eye-catching appearance which demanded attention regardless of quality. Each had an individual personality, and while the precision which its advertising claimed for it was present to some degree, it often lacked the suggestion of superior craftsmanship, the heft and feel of quality merchandise. Many were smooth and functional in appearance but proved bulky and

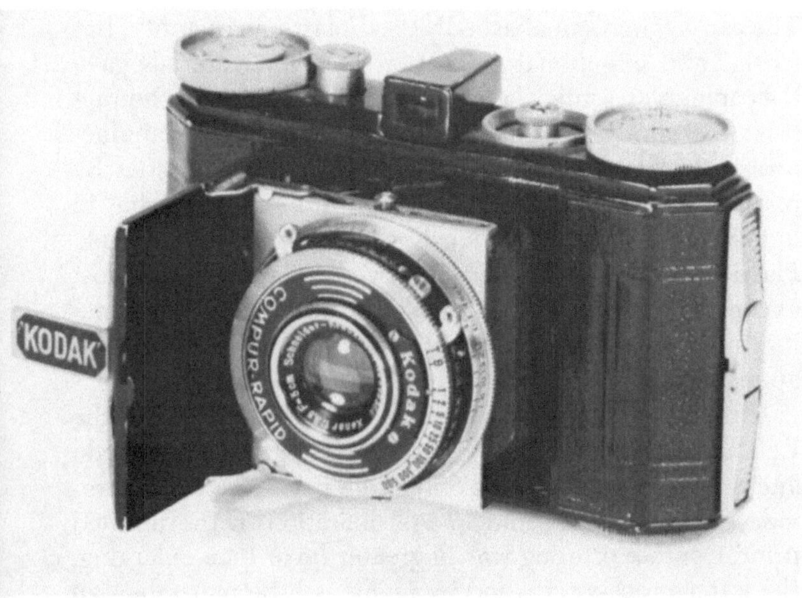

Although manufactured in Kodak's German plant at Stuttgart, the Retina I was important to the development of the American 35mm for two reasons: it represented Kodak's entrance into the 35mm market, and, perhaps more important, it introduced the Kodak 135 magazine. This packaging of 35mm film in a standard daylight-loading container made it possible for other manufacturers to design cameras without worrying about developing their own film-handling systems.

awkward in use, making it necessary for the users to concentrate on their operation instead of their pictures.

Such problems could sometimes be attributed to the camera's construction. Three kinds of construction were generally used, die casting, stamped metal, and molded plastic. Occasionally all three were combined in one way or another.

The early 35mm's unabashedly used plastic in quantity, but its feel and intentional simulation of other materials gave the cameras an artificial appearance which tended to detract from the aura of quality. In later years camera manufacturers went to great lengths to disguise many plastics by plating them or using a leather or leatherette covering to hide what the public obviously regarded as a cheap look. Plastics won out in the end, however; today even the most expensive cameras have exposed plastic parts which somehow manage to look appropriate and even add to the quality appeal.

Until the 35mm miniature appeared in the form of the Leica, it was an unwritten but generally agreed-upon credo among photographers that only large negatives would provide sufficient detail and image definition in the finished print. Contact printing was in greater favor than enlarging, the latter process regarded by many as little more than an amplification on paper of the poor image quality of the average lens. Thus the key to the success of the Leica—and the 35mm movement—in overcoming the supposed drawback of its small negative was in no small measure the lens.

Looking back today at the large number of trade names which appeared on 35mm optics during the period, the reader may be surprised to learn that a high percentage of them were actually design variations of two popular formulas, the Cooke triplet and the Zeiss Tessar lenses. Even the famed Leitz Elmar *f*/3.5 lens was a variation of the two. While a treatise on lens design and formulas is not appropriate to this book—it is a far too complex subject—a few general observations are in order here to establish a perspective on the major attributes of the 35mm miniature camera.

A single lens, like a magnifying glass, can be used as a camera lens. The image it records on the film during the exposure is very fuzzy owing to the manner in which light rays that form the image are bent by the lens surface. At very small apertures (up to about *f*/11), specially designed single lenses make fairly sharp pictures. Most simple box cameras contain single-element lenses. Unsharp images of single lenses used at larger apertures can be traced to specific ways various light rays pass through the lens. Though the total image is always the sum of the light rays, the lens designer breaks down the various specific causes of unsharpness into categories called *aberrations.* Unsharpness in the center of the lens field caused by a difference in the focal length of the center and of the edges of the lens is called *spherical aberration.* A similar aberration in the field is called *coma. Astigmatism* is another aberration that causes unsharp images in the field area of the image. By combining a number of lenses (lens elements) of different shapes and types of glass into one lens assembly, the manufacturer makes a *corrected lens* which, because one lens element compensates for aberrations in the others, produces a sharp image in which the effects of the aberrations are minimal.

Designed in 1893 by H. Dennis Taylor, an Englishman, the Cooke triplet was among the first fully corrected, large-aperture lenses which also possessed a flat field. While low-priced triplets were usually inferior to other designs in performance, possessing a rather noticeable loss of edge definition compared with their excellent center definition, they were often used on the less-expensive 35mm's, providing a sufficiently good negative to satisfy the needs of their owners. The Tessar lens was a standard four-element lens (three-element if one considers the two rear elements which are cemented together as one, and some experts do). It was derived from the triplet in 1902 by Dr. Paul Rudolph, of the Zeiss works in Jena, Germany. It incorporated the best features of many earlier optics, offering better spherical and color correction than most comparable lenses.

When the Zeiss patents expired in 1922, every manufacturer (or so it seemed) rushed from the designing board to market his own variation. For moderately high speed (*f*/2.8) and good correction there was (and is) no equal to the Tessar formula, and modifications appeared containing as many as seven elements.[1] Expensive and highly corrected lenses like the Xenon, Biotar, Summar, and Summitar (all *f*/2 and of foreign manufacture) were also available with Gaussian-type objectives of virtually equal sharpness from center to edge. But most 35mm cameras falling within the scope of this book used less than perfect variations of either the Cooke or the Zeiss formulas.

Designers of miniature cameras early settled upon the 50mm focal length, which gives a slight telephoto effect. A focal length of 44mm is more nearly normal in length necessary to cover the diagonal of the 24 × 36mm negative. The relationship between maximum lens aperture and field angle seems to have been the probable reason for the adoption of 50mm as "normal." It is much easier to design relatively large aperture lenses that cover small field angles than it is to design them to cover large field angles. Even today, *f*/1.4 wide-angle lenses are exceedingly rare. By keeping the focal length somewhat larger than the diagonal of the film format,

[1] At various stages of lens technology *f*/4.5 and then *f*/3.5 were the top lens speeds attainable with the Tessar formula. With the advent of rare-earth glass (made from relatively rare minerals instead of sand, as was all early optical glass and most glass even now), excellent quality *f*/2.8 Tessar-type lenses became possible.

it was possible to manufacture the "sharper" lenses required by the small film format with the larger apertures that made candid photography possible.

While the properties of antireflection lens coatings were known as early as 1892, modern coating dates from a 1935 Zeiss patent. As light is transmitted through the various lens elements en route to the film, internal glare and reflections are created. The application of a thin (1/200,000-inch) transparent coating, usually a metallic fluoride, to the lens served to minimize this problem, providing the lens with vastly improved contrast and at the same time slightly increasing its effective speed. Although a few expensive 35mm models like the Kodak Ektra camera pioneered in the technique shortly before the outbreak of World War II, coated lenses were not in common use until after the war.

The early coatings were "soft" and thus easily damaged or removed. For that reason they were applied only to the inner surfaces of the lenses. Wartime research to extend the marginal visibility of lenses under adverse conditions led to accelerated development of the process, including "hard" coatings that resisted damage and removal. All but the most inexpensive postwar cameras appeared with bluish, purplish, or brownish coated lenses.

The shutter design of the 35mm camera was also important to its success. The leaf, or blade, shutter and the focal-plane shutter were, and continue to be, the two major kinds. The Ilex Universal (leaf) and the Leica (focal-plane) shutters provided the prototypes for most shutters of the period.

The leaf shutter is composed of a series of pie-wedge metal blades that converge to form a whole. When the shutter is released, the blades, or wedges, pivot out of the way for the duration of the exposure. A typical focal-plane shutter consists of a length of rubberized cloth or metal on rollers which incorporates a pair of flexible but opaque "blinds" that traverse the film's face. Releasing the shutter rolls the curtain across the film, and the second blind follows the first after a fractional interval, creating a slit which exposes the film a portion at a time. The amount of light reaching the film is determined by one of three methods: a fixed speed of travel combined with a variable slit (Leica), a variable speed of travel combined with a fixed slit, or both variable speed of travel and variable slit (Contax). The curtain must be rewound after each exposure; in 35mm cameras this operation is usually coupled to the film-wind mechanism. Variations in focal-plane design included the UniveX Mercury's rotary focal-plane, a circular piece of metal with a variable window which was revolved in front of the film at a constant speed, the window size determining the actual exposure speed. Today the latest focal-plane-shutter technology seems to be forecast in the unit-built Copal Square all-metal format.

One of the major problems faced by designers was a lack of space for the shutter. The most common placement of the leaf shutter was between the lens elements, but with the smaller size of the 35mm-camera lens the amount of space available to incorporate it was correspondingly reduced. Many designers placed their leaf shutters behind the lens or chose to use a focal-plane shutter. In either case the shutter could then be built into the camera body, allowing much more room for placement of the intricate mechanisms necessary to drive and time the shutter. That possibility also reduced the complexities of lens-mount design and allowed for interchangeable lenses. Such shutters permitted a rugged design, as evidenced by the Argus C-3 Micromatic (behind-the-lens) which was sturdier than the Ilex Precise or the Wollensak Alphax (both between-the-lens). Relatively easy to design and manufacture, the focal-plane shutter was even more rugged and its accuracy considerably more efficient than the simpler leaf shutters, especially at higher speeds. The major concern rested in the choice of shutter speeds. Generally speaking, the focal-plane shutter could be made to give faster speeds (to 1/1250 at the time) with a comparatively simple design; when speeds slower than 1/20 second were desired, the problems began to appear.

The right-to-left travel of the Leica cloth shutter was a much more desirable design to duplicate than the metal-leaved, up-to-down Contax shutter (which also possessed the annoying habit of shearing its cotton ribbons in half in those days before nylon).[2] But when American manufacturers of cameras like the Perfex, Zephyr, and Detrola adapted the Leica shutter to avoid charges of patent infringement, they usually produced shutters which failed to work properly, consistently, and accurately.

American leaf shutters were interesting configurations of gears, springs, cams, and other exotic contrivances. Most of the popular ones (such as the Kodak Supermatic and the Wollensak Rapax) were essentially variations of the German Deckel Compur, which appeared in 1912 with the same gear-and-detent arrangement patented by Rudolph Klein and Theodor Bruech of Ilex in their 1910 Universal shutter.

[2] Leitz bought its shutter material from the American manufacturers of Graflex cameras until the outbreak of World War II. It was only through Graflex's cooperation with the United States government that Leitz could resume production after the war, and a few rare postwar Leicas have the red-and-black shutter blinds which resulted from this agreement.

Klein and Bruech were onetime employees of Bausch & Lomb and designed many shutters for that company, including the famed Volute in 1902. Later they organized the XL Manufacturing Company to manufacture shutters and on June 25, 1910, applied for Patent No. 1,092,110, which was assigned to a wholesale jeweler, Morris Rosenbloom, who apparently financed them. A year later, XL changed its name to Ilex Optical Company, and with Rosenbloom's management talents and the designs of Klein and Bruech, Ilex quickly became a major American shutter manufacturer, a position it held until Rosenbloom's death. After that the company gradually lost ground until the early sixties, when it was purchased by Manuel Kliner and Eugene Miller (former employees of Elgeet, another Rochester optical company, now defunct). Thriving once more, Ilex is the only remaining manufacturer of large shutters in the United States.

After World War I camera manufacturers began indicating that they would like to buy shutter assemblies complete with lenses. In 1922 the Ilex Optical Company absorbed the Acme Optical Company, and in the thirties its contribution in leaf-shutter design to the American 35mm was everywhere in evidence. Particularly outstanding was its very inexpensive Precise shutter, manufactured in two sizes (OO and O), one of which graced the Model A Argus in 1936. The Ilex lenses and shutters were used mainly by Argus and Universal in the 35mm field, and the company's influence peaked somewhat earlier than that of its rival, Wollensak, which after World War II became the major supplier of lens-and-shutter combinations for independent makers of low- and medium-priced 35mm cameras.

The Compur shutter, based on the Ilex patents (for which

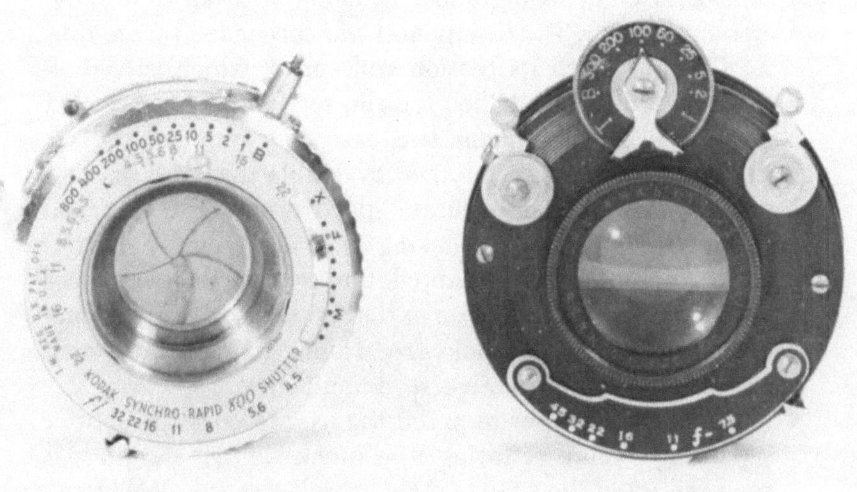

Two types of leaf shutters are the rim-set (left) and the dial-set (right). The Kodak Synchro-Rapid 800, which utilizes rotary blades, represents the ultimate in shutter development in the United States. The early Ilex is typical of the shutters available before the development of rim-set models.

Deckel paid Ilex a royalty on a per-shutter basis), was a great improvement over Deckel's earlier Compound shutter, which had used an air-retard system to control exposure time. While the Compound was an excellent shutter, it had certain shortcomings common to all air-retard shutters and tended to be erratic in operation. Early Compurs were dial-set: the shutter selector dial replaced the Compound's air cylinder on the top face of the shutter. Rotating the dial moved a series of cams which controlled the retard mechanism. The great number of parts in a small space caused considerable jar at high speeds, along with an inordinate amount of wear. The Compound was succeeded by the rim-set Compur, with its tension coil spring which served to lessen jar and wear. When the rim-set Compur was scaled down to fit into the 50mm lens used on 35mm cameras, its shutter blades were reinforced to double thickness at their pivot points, and an accurate speed range of 1 to 1/500 second became possible, making it the ideal quality shutter. Other manufacturers adapted the escapement-retard, or gear-detent, system utilized in the rim-set Compur, producing a number of variations free from patent infringement and relatively inexpensive to manufacture.

While the Compur required the user to cock the shutter separately before releasing it, a much simpler design was represented in the single-action, or self-cocking, Wollensak Alphax; one movement of the release lever both set and tripped the shutter. The Alphax was a lever-set rather than a rim-set shutter; a lever moved around the top of the shutter and was set to specific markings. While its speed range was restricted to 1/10 to 1/200 second (an absolute limit—it usually ranged between 1/25 and 1/150), it became the most common of the leaf shutters used in inexpensive American 35mm cameras after World War II.

One of the famous "job shops" to the American optical industry, Wollensak was founded in 1899 by a German immigrant, Andrew Wollensak, who left the Bausch & Lomb Optical Company to manufacture camera shutters on his own. The venture succeeded so well that by 1902, when his customers inquired about the possibility of lenses, he and his brother, John Wollensak, united to produce the first Wollensak lens, a single achromatic selling for 75 cents.[3] Thus began an association which eventually led to Wollensak's dominant position in the production of medium-priced lens-and-shutter combinations, despite the fact that the rejection rate was often as high as 60 per cent during the best years. The Wollensak Alphax (to which flash was later added as the Synchro-Alphax) was a very popular shutter in the low-to-medium-priced 35mm cameras, especially after the war, when manufacturers bought the shutter-lens units in huge quantities. The Perfex One-O-One, Perfex One-O-Two, Cee-ay, Ciro 35, Graphic 35, Bolsey B, Bolsey B2, Bolsey C, and many others in the under-$100 price bracket featured this combination.

In spite of its limitations the Wollensak triplet lens satisfied its users' demands and was usually found in focal lengths of 44 to 50mm, with a top speed not exceeding *f*/3.2 Simplicity in itself, compared to Compurs and Compur Rapids, the accuracy of the Alphax self-cocking shutter varied somewhat, especially after it had been used for some time, but it appeared to be as satisfactory to owners of the cameras using it as the accompanying lens, and fewer of these units came to the repair bench than more exotic designs. Of course, many owners never realized that their camera's shutter was at variance with its marked speeds and simply adjusted by selecting combinations of shutter speed and lens opening which compensated for the deficiency.

Many of the early 35mm miniatures were focused manually; the user rotated either the front lens element (a usual feature of a triplet lens) or the entire lens assembly

[3] The first print made from the initial exposure taken with the first lens of this series was preserved under glass for decades at Wollensak, and as late as 1968 still exhibited a photographic quality which must have appeared excellent to users of its day.

MINICAM MECHANICS

When lens is focussed sharply upon point P, rays from P converge upon the film at P'. Optical properties of lenses, however, cause the image to be not a point but a disc, whose diameter is represented by the line X-Y. Any other points whose rays fall upon the film within this disc will also be in focus and the distance between the extreme ones – P_1 and P_2 will be the DEPTH of FOCUS.

The focussing range of this finder is from 3 feet to 700 feet. Objects nearer than 3 feet cannot be put in focus without special equipment. Objects beyond 700 feet are at "infinity" and are in focus.

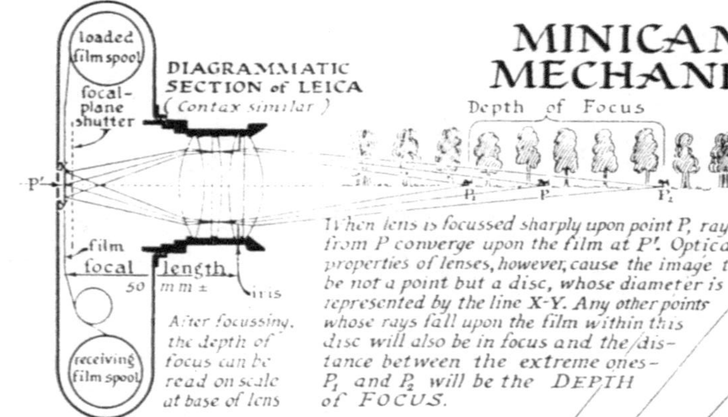

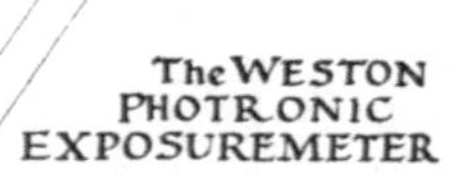

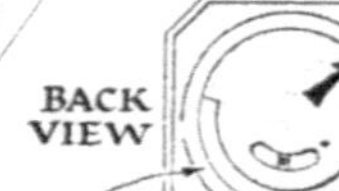

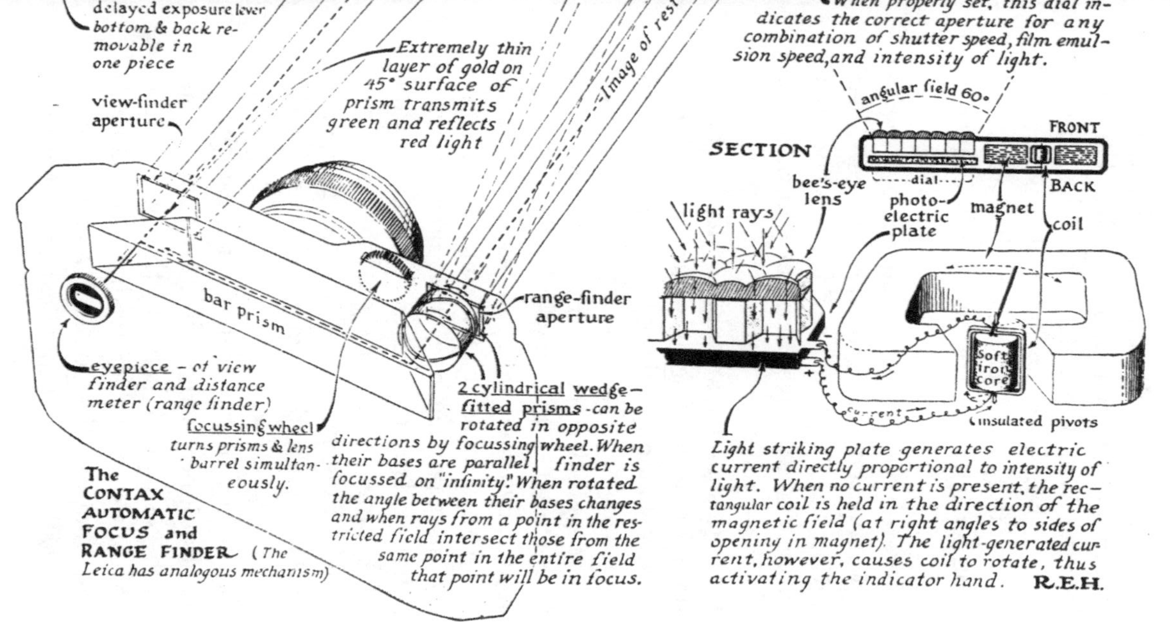

Reproduced through the courtesy of Fortune *magazine.*

in a helical mount. But rangefinder focusing soon became a standard feature on most 35mm cameras offered to the public, especially after World War II. When the camera did not incorporate a built-in rangefinder, it was usually available as an accessory (or at least an attached accessory clip made provision for it), as in the case of the Universal camera line.

Rangefinders were of two types, the split-image "military" and the superimposed image, or coincidence. Their principle of operation was simply to provide two views of a selected object from slightly separated viewpoints, one a direct line-of-sight view, the other passing through a semi-transparent mirror or prism set at an angle 45 degrees to the axis. The function of these focusing aids was usually found in a separate viewing window, although in the combined rangefinder and viewfinder, which became very popular after the war, the rangefinder occupied a small rectangle, circle, or triangle centered in the viewfinder field. One or both images were usually tinted with contrasting colors for clarity of view and ease in focusing. The coincidence viewfinder became the most popular one; most users found it easier to match the overlapping images. Exceptions were the postwar Kodak rangefinders, designed by an expert in military split-field rangefinders who produced highly sophisticated and reliable finders, such as that used in the Kodak Ektra camera.

These were the main features of the American 35mm miniature camera. Equipped with such instruments, thousands of photographers snapped their way through the thirties and forties. While we have given credit to the camera's lens as a major attribute in the popularity of the 35mm movement, that is really an oversimplification which must now be amended, for without film capable of reproducing the best that a lens could offer, the lens would have been merely a bright, shiny, but useless piece of glass, and the 35mm camera would have remained a toy of the wealthy.

4 From Black and White to Living Color

Film as we know it today began as an idea produced by George Eastman's fertile imagination. Eastman, an amateur photographer, began making his own photographic emulsions in 1878. Six years later he was manufacturing Eastman Negative Paper, a light-sensitive emulsion on paper from which prints could be made by treating it with hot castor oil after development to effect a degree of transparency. Eastman American film, a stripping film, followed the next year. After development the paper was removed, or stripped, leaving a thin negative film which was then mounted on glass for printing.

In 1889, eleven years after he started, Eastman was able to announce the first commercial transparent roll film. He had perfected a properly coated, thin, flexible sheet of cellulose nitrate with a uniform gelatin emulsion containing suspended microscopic grains of light-sensitive silver salts. By substituting film for glass plates commonly used in the 1880's, he had provided the key to an eventual billion-dollar industry. Through his driving determination for consistency, his faith in basic as well as applied research, and, of course, his resulting patents, Eastman led from the beginning in the manufacture and sale of sensitized products of the highest quality.

Before the appearance of the Leica small-format cameras failed to achieve any significant degree of popularity, mainly because of the lack of film adequate for their use. The films contemporary with these cameras were "slow," or comparatively insensitive to light and thus restricted the camera's use to bright, sunny days, in effect negating its very reason for existence. Although new, faster types of emulsions were produced, they lacked the high definition needed for enlargement, giving a grainy appearance to larger prints, and this shortcoming severely restricted their use with small cameras. When the Leica appeared in 1925, the fastest black-and-white films available were comparable to today's Panatomic X film in speed.

Film evolved from a slow emulsion sensitive only to the blue rays of the spectrum to the faster orthochromatic emulsion with its increased sensitivity to blue and green, and finally to the panchromatic emulsions sensitive to all three of the additive primary colors, red, blue, and green. Along with the increase in color sensitivity, the effective speed of films was also increased.

Other factors to be considered in producing an emulsion were its grain structure, exposure latitude, and tonal gradation.[1] It was by no means an easy job to make a new emulsion that satisfied all the requirements of a film for 35mm cameras, and years of exacting laboratory research went into each new film. But thanks to many advances in color

[1] *Gradation* is the ability of the film to reproduce in a natural manner the many variations of black and white present in the subject being photographed. As a general rule, slow-speed films have a *hard gradation*—the light-and-dark contrast appears greater than in the subject; high-speed films have a *soft gradation*—there are more grays and fewer sharp blacks or whites. The gradation of the final negative is dependent upon two factors: the nature of the emulsion and the degree of development.

sensitizers, an increase in the emulsion thickness, and the use of gold salts as speed sensitizers, film improvements accelerated in the early thirties. With film which could satisfactorily record what the high-quality lenses could see, the 35mm camera users were off and running.

Kodak's closest competitor was Ansco (today the GAF Corporation). Its genealogy can be traced to Edward Anthony, who is generally considered the first American photographic dealer. Anthony opened a daguerreotype supply house at 308 Broadway in New York City in 1842, and some years later his brother, Henry T. Anthony, joined him in founding E. & H. T. Anthony. In 1902 the firm merged with Scovill and Adams Company. The resulting unwieldy name was contracted to Ansco in 1907. Another merger took place in 1928, this time with the American interests of Germany's Agfa film organization, and the name was changed to Agfa Ansco.

But Ansco got a late start in production and, with a somewhat provincial attitude toward advertising, soon found itself in an unenviable position: Kodak led, and it could only follow. Almost without exception Ansco was forced to find ways around Kodak's patents, problems which always took time to solve even though sometimes the issue was as simple as avoiding the film-numbering system which Eastman had patented.[2] While Ansco was spending valuable time resolving such problems, Kodak's research, its production lines, its strict quality control, and its familiar yellow box were easily capturing the market.

[2] While we take film-size designations for granted today, Kodak devised the system and protected it by law for many years. For example, a roll of Kodak film was designated as 120, but Ansco was unable to use this system, and its identical film was designated B2, a source of some confusion to the consumer.

Other competitors faced the same disadvantages. Du Pont Film Manufacturing Corporation, Inc., also manufactured film, but Du Pont was never really a serious contender in the amateur field. The only significant foreign manufacturers to try to penetrate the American market were England's Dufaycolor and Belgium's Gevaert. The success of the latter was achieved not under its own name but in private-brand packaging.

For several years after their introduction both the Leica and the Contax cameras made use of specially constructed reusable film containers. Precision-machined of brass, these film magazines, or "cassettes," as they were popularly called, were designed for home loading of film in either a darkroom or a changing bag. At first, users bought rolls of 35mm nitrate movie film and cut off a length sufficient to fill the cassettes (5¼ feet for 36 exposures), but when prepackaged lengths were made available by film manufacturers, this practice became less popular, and the mail-order reloaders took over.[3] When the cassette was inserted and the camera's base plate locked, a cam on the plate activated the cassette pin, lifting the safety spring and opening a slit, or door, about ¼ inch wide, allowing the film to pass through it. When the roll was rewound and the base plate unlocked,

[3] Respooled film was very popular until the late forties. Many of today's largest dealers started as "short-end" loaders—a somewhat hazardous occupation, since much of the film had a nitrate base which had a propensity toward deterioration and was also highly combustible. Probably the most famous short-end dealers were Harry and Ben Teitelbaum, of Hollywood Film Company, and Harold Rosen, who with his brother started Minifilm Labs (a processing service) and later expanded into the hardware end of photography with Minifilm Camera, a large New York mail-order shop still operating today. Short-end loaders bought odd lots of bulk 35mm movie film and spooled it in used cartridges, a practice which all but ended when Kodak introduced the crimped-end cartridge, which cannot be reloaded.

the safety spring closed the slit and prevented light from reaching the film.

The Contax required two cassettes, as did the Kine Exacta, another German 35mm, made by the Ihagee Camera Company. One cassette held the unexposed film, and the other received it after exposure, permitting removal of the film in daylight without rewinding. The Leica user had to rewind the film into its original cassette before removing it. Adherents of the two systems argued their respective merits for years, with theory resting firmly on the side of the Contax. Dust combined with movement of the film across a surface can lead to film scratching, and the fewer times a 35mm film is wound the less likely the chance of inadvertent scratching.

Preparation of the film cassettes in a darkroom, as well as the expense to those who wished to preload several at a time, imposed a severe handicap on the hobbyist. Had this limitation not been overcome, there is good reason to suspect that the 35mm format would have failed to rise beyond the favor of a small, dedicated group.

When the Leica made its formal debut on the American scene in 1932, Kodak's most popular emulsion, Kodak NC (Non-Curling) film (which had been its standard film for amateurs since 1903) had given way to the new Verichrome film, but neither emulsion ever appeared in 35mm. Kodak's first panchromatic film, Super Sensitive Panchromatic (SS Pan), made its debut in November, 1932, in both rolls and 35mm loads of 36 exposures. An SS Pan load designed for use with the Contax cassette sold for 90 cents; a Leica cassette load of 30 exposures was marketed in June, 1933, at 70 cents. Each had a strip of paper attached to the film which served as a leader, simplifying the loading process.

In the same month Kodak's Panatomic film, the first fine-grain emulsion designed especially for 35mm use, appeared in Contax loads at 85 cents each; the cost of a Leica load remained at 70 cents.

About the same time Agfa introduced several types of Leica loads in cardboard containers which replaced cassettes. Although they were an innovative attempt at making loading easier, they were abandoned when Kodak introduced the daylight-loading cartridge in December, 1934, with the Kodak Retina camera.[4] For the first time anyone with a reasonable degree of dexterity could easily load, unload, and reload a 35mm camera in broad daylight without ruining the film. This packaging innovation made it possible for American camera manufacturers to enter the 35mm market with inexpensive products like the Argus line. The small format now became popular with large numbers of consumers, and in the next few years many kinds of films appeared in 35mm (see Table 1), in various lengths ranging from 10 to 36 exposures. The economy in cost and space made 36-exposure loads the most popular. Although companion loads providing 18 exposures appeared in mid-1939, the shorter length did not really catch on until after World War II, when manufacturers arbitrarily extended it to 20 exposures for those amateurs wishing a compromise length to hasten completion of a roll. These two lengths remain

[4] The history of the 35mm film magazine is one that the authors have not yet resolved. While it appears relatively certain that Kodak was the first user of what is today's basic 35mm film holder, it is not known just where the idea came from. Even the company files and publications do not shed much light on the matter. The film holder was introduced with the Retina camera and termed a "magazine." However, later sales literature referred to it as a "magazine cassette." The Leica and the Contax reloadable film holders had been called cassettes, and possibly this change in

standard today, and repeated attempts to market shorter loads have met with little success.

Popular acceptance of the 35mm format spurred film researchers in other directions. At first standard motion-picture stock was the sole film available for picture taking (Perutz brought out its Leica Special film in 1929), and although many millions of feet of movie film had been manufactured by 1925, companies had found it difficult to produce a length of 35mm film without imperfections, owing to the small size of the film and the manufacturing techniques then in use. As far as motion pictures were concerned, such flaws were not a serious problem.

With the advent of 35mm strict quality control of still emulsions became the order of the day, and each film manufacturer had its own research laboratory, where all kinds of unusual problems turned up for study and analysis. The work of Kodak's Robert N. Titus and his staff was typical. Microanalysis of photographic emulsions became routine lab procedure, but occasionally particularly difficult problems arose which plagued assembly-line production or threatened quality control. For example, isolated specks of an impurity which could not be detected by ordinary procedures might appear in the emulsion, invisible until the negative had been enlarged many times its original size.

terminology was made to inform salespeople that the Kodak magazine would work in those cameras. But the plot thickens: other literature refers to 35mm film "retorts" for the Retina camera. One meaning of "retort" is a container from which something can be removed and replaced. Our guess is that the word was used in Europe, but we have yet to find proof. It is also interesting to note that in the early days George Eastman called his roll-film cameras "cartridge loading"—a term that somewhat preceded the Kodak Instamatic camera, which is also cartridge loading.

Table 1

Available Films and Weston Speed Ratings, 1938–39

Compiled from various sources, the following list of 35mm films includes the most popular ones available in the United States in 1938–39. Foreign films, such as Selo, Mimosa, and Cappelli, could be purchased in larger photographic stores. The period chosen clearly illustrates two aspects of the era: the ever-changing speed ratings assigned by Weston (see Chapter 6) and the beginning of the great leap forward in superspeed black-and-white emulsions in 1939. To give the reader an idea of the film's intended use or its property, a phrase from the manufacturer's description is given after each kind of film.

Film	Use or property	Type	*Weston rating 1938* Daylight	Tungsten	*Weston rating 1939* Daylight	Tungsten
AGFA ANSCO						
Ultra Speed Pan	Extremely fast	Pan	64	40	125	64
Fine Grain Super Pan[a]	Very fast	Pan	24	16	64	40
Finopan	Very fine grain	Pan	16	10	24	16
Plenachrome	Fine grain	Ortho	16	6	24	12
Infra-red	Haze cutting	IR	2[b]	—	2[b]	—
DUFAYCOLOR						
Dufaycolor	Additive screen-type color	Color	8	3[b]	8	3[b]
DU PONT						
XL Pan[c]	Very fast	Pan	—	—	64	40
Superior Pan	Very fast	Pan	24	16	40	24
Fine Grain Parpan	Fine grain	Pan	8	4	10	8
Micropan	Very fine grain	Pan	4	3	6	4
Infra-D	Haze cutting	IR	12	6	12	12

GEVAERT						
Panchromosa	Very fast	Pan	24	16	32	20
Panchromosa Microgran	Very fine grain	Pan	12	6	12	8
Express Superchrome	Fast ortho	Ortho	20	12	16	12
KODAK						
Super X Pan[d]	Very fast	Pan	24	16	100	80
Super Sensitive Pan[e]	Fast fine grain	Pan	20	12	40	24
Panatomic[f]	Very fine grain	Pan	16	10	32	20
Kodachrome Regular	Subtractive dyed image, screenless	Color	8	3[g]	8	3[g]
Kodachrome Type A	Subtractive dyed image, screenless	Color	8[g]	12	8[g]	12
Infra Red	Haze cutting	IR	4	6	4	6
PERUTZ						
Peromnia	Very fast	Pan	24	16	16	10
Perpantic	Fast fine grain	Pan	20	10	4	2
Pergrano	Very fine grain	Pan	10	6	5	3
Neo-Persenso	Fast fine grain	Ortho	20	8	4	1.5
UNIVEX						
Ultrapan	For adverse light	Pan	16	12	24	12
Ultrachrome	For average weather	Ortho	16	6	24	10

[a] Replaced by Superpan Supreme in 1939.
[b] With appropriate filter.
[c] Introduced in 1939.
[d] Replaced by Super XX.
[e] Replaced by Plus X.
[f] Replaced by Panatomic X.
[g] With appropriate filter.

When such impurities showed up during emulsion testing, Titus stepped in and, using microtools of his own design and manufacture, isolated and tested the specks to determine their origin. Men like Titus were unsung heroes of the 35mm movement.

While the development of satisfactory 35mm films slightly preceded the advent of the American 35mm camera, another emulsion advance was just around the corner which held untold implications for the future of the miniature camera. Attempts at color photography date back to the beginning of the science, but color received its greatest impetus from the work of physicist James Clerk Maxwell, who in 1861 demonstrated the additive process by superimposing on a screen the projected images of three identical lantern slides (one red, one blue, and one green), projecting each through a filter of the same color used to make the reproduction.

Many researchers followed Clerk Maxwell, but few processes reached the commercial stage. One such, the English Dufaycolor, was available to the minicam fan and could be developed at home, but it contained a mosaic filter pattern which was visible when viewed and objectionable when enlarged by projection. It was also a "dense" film, and the projected images were much dimmer than the black-and-white slides popular at the time because they contained both a silver image and the mosaic pattern. Kodachrome and later color films achieved color by changing silver images into dye images, eliminating the need for the mosaic pattern and producing slides as brilliant as the best black-and-white ones had been. But color photography lingered offstage until 1935, remaining a province of a few professionals using cumbersome equipment and complicated

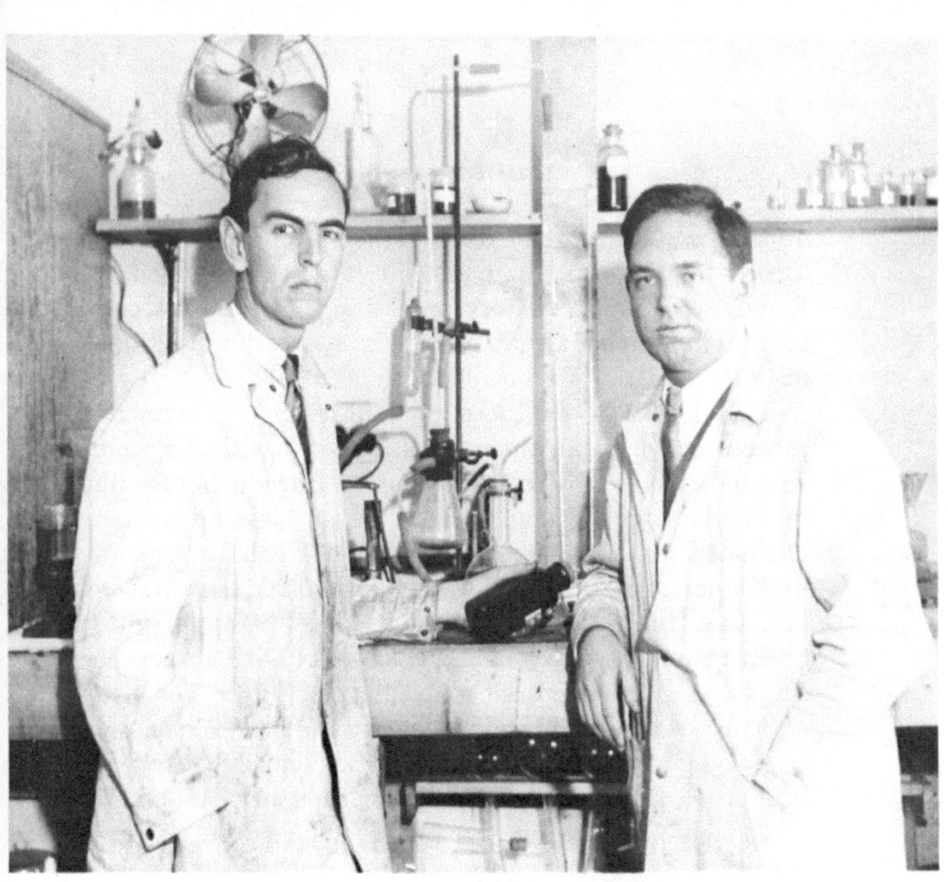

Two musicians, Leopold Mannes (left) and Leopold Godowsky, shown in what has become a classic press-release photograph. From their "kitchen chemistry" came the basic process for Kodachrome. When they took their discovery to Dr. Kenneth Mees, of Kodak Research Laboratories, the company was on the verge of introducing a two-color film. The project was set aside, and development of Kodachrome was given top priority. The rest, as they say, is history.

processes, until Kodak introduced its 16mm Kodachrome film to home movie fans in May of that year.

Basically the idea of Leopold Godowsky and Leopold Mannes, two young photographers (who were also famous musicians), Kodachrome dates to their attempts to produce a color film in 1921. In 1930 the two inventors joined the Kodak Research Laboratory, and by 1934 their efforts, combined with those of Dr. Mees, were so successful that Kodak decided to market the new film commercially.

Kodachrome is produced by a three-color subtractive process in which all three emulsions rest within a single coating separated by layers of gelatine. Introducing the film required new processing facilities, tons of special processing chemicals, and hundreds of thousands of labels, cartons, and instruction sheets. A lot of work was involved, but in October, 1936, the new color film made its debut, appearing in both 35mm (K135) and 828 Bantam (K828) size. The K135 was an 18-exposure roll selling for $3.50. The K828 was an 8-exposure roll and at first sold for $1.75. The cost of processing and return postage was included in both prices. The processed film was returned in strips, unmounted.

Kodachrome film was an instant success, but it really came into its own in 1939, when do-it-yourself slide mounting went by the board and the laboratory began mounting individual transparencies in the familiar 2 × 2-inch cardboard Readymounts still with us (although purists continued to hand-mount their transparencies in glass for projection). The film came in daylight (Regular) and indoor (Type A) emulsions rated at Weston 8 and 12 respectively. It opened up an entirely new and profitable range of accessories—conversion filters, slide-cover glass, masks, tape, cardboard mounts, hand viewers, projectors, screens, and so on.

Acceptance of the new color film by professionals and nationwide exhibition of their work whetted the appetite of 35mm miniature camera fans and others. How many 35mm cameras were sold because of color film is impossible to de-

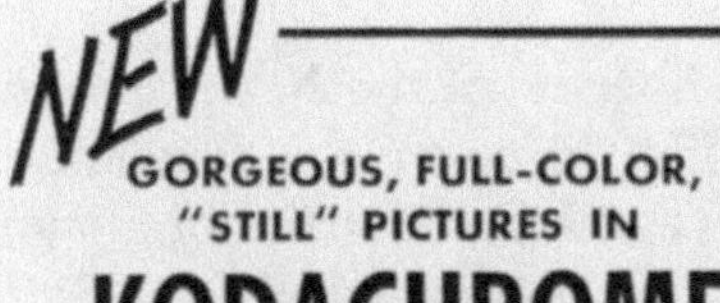

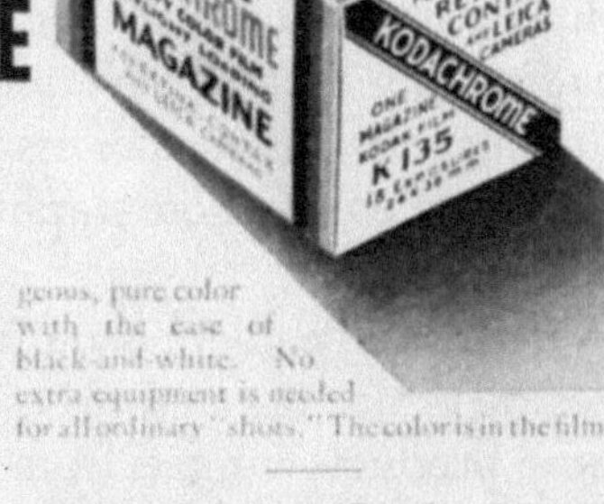

This 1936 advertisement introduced the "new" Kodachrome color film. Kodak had developed a two-color process in the early 1920's but had decided not to market it. There was also a version, introduced in 1934, made by a lenticular process and released only in 16mm. Actually a black-and-white film, it was exposed through and projected with a special filter system to produce the color image. Since it was black and white, it never faded and had few of the processing problems of today's color; however, it required a great amount of light and was both impractical and unpopular.

termine, but an educated guess would claim a substantial number. Only a short time after its introduction Kodachrome found its way into newspaper rotogravure sections. Robert Dumke's Contax transparencies of Victor Herbert's operetta *Sweethearts* appeared in the *Milwaukee Journal* on January 10, 1937. Gerard Sheedy, of the *New York Sunday Mirror*, used the new film to cover the *Hindenburg* disaster in May of the same year. Ivan Dmitri shot a 35mm color transparency that was used on the cover of the *Saturday Evening Post*. The *National Geographic Magazine* published its first 35mm color essay in 1938, and by 1939, 90 per cent of its color photographs were being taken on 35mm film.

Virtually all camera manufacturers brought out their own lines of projectors and accessories for color-slide viewing. These products ranged from simple, hand-held boxes with a slot for the slide, which was to be pointed at a light source for illumination, to elaborate desk models equipped with a lens, a built-in light source, and opal-glass diffusion. The Society for Visual Education profited from its lengthy association with filmstrips and the educational world. SVE, as it is still known, had produced an extensive library of black-and-white educational filmstrips over the years and sold a series of filmstrip projectors which were easily adapted for slides. Primitive in design but reliable, the SVE projectors were part and parcel of the 35mm color scene during the thirties and forties, under either the SVE banner or that of private labels like Argus, for whom they manufactured a line sold as the Model B, the first of many Argus projectors. Spencer Lens Company (later known as American Optical) joined the picture in the same way, manufacturing the DP series for Argus and selling under their own name at the same time. Eastman Kodak Company, which had started it all, introduced the first of a long line of distinguished Kodaslide projectors, and new firms came along, among them Vokar (which made Sears, Roebuck's first projectors), TDC, La Belle, Golde, and dozens of others.

Projectors evolved slowly in the prewar period, and blower cooling units and push-through slide carriers were about the only exotic improvements, while designs ranged from ugly to very ugly. After World War II the Airequipt and, almost simultaneously, the TDC changer were introduced, and the rush was on toward automatic slide projection.

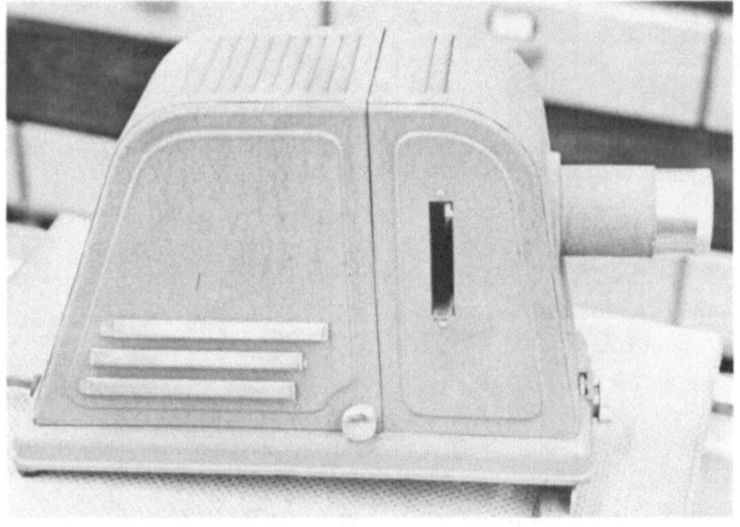

The Argus Model DP slide projector minus the push-pull slide carrier which fit into the empty slot in the front panel. Identical to the Spencer Model MK Delineascope (both made by the same company and sold to the 35mm enthusiast at $22.50 in the late thirties), this projector was convection-cooled and came equipped with a 5-inch f/3.6 lens and 100 watts of illumination. It weighed a solid 5 pounds.

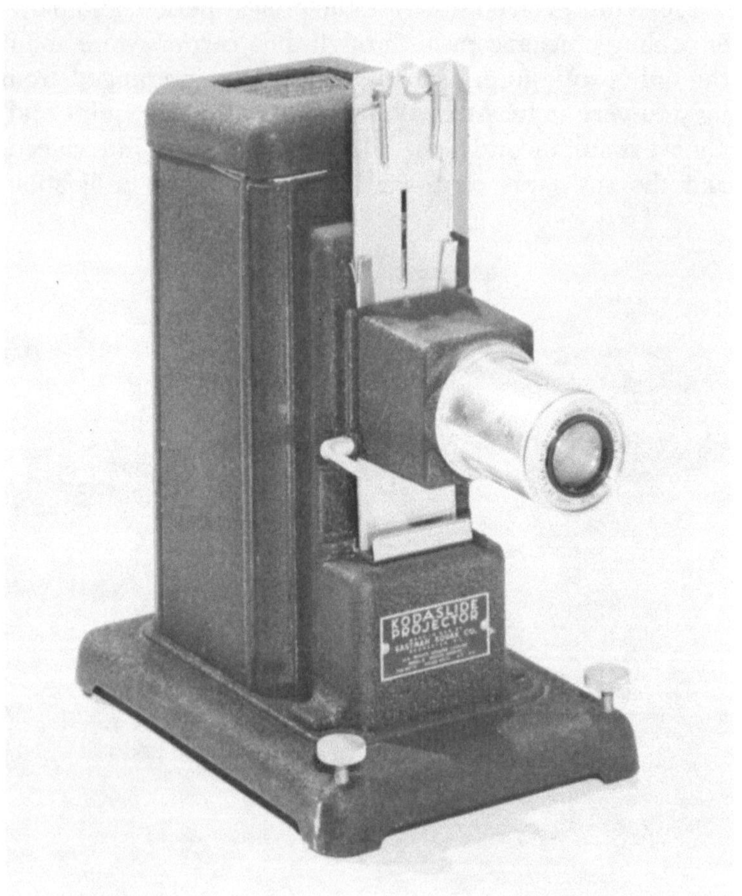

The Kodaslide, the first Kodak 2×2-inch slide projector (shown with an accessory slide carrier), set a pattern of styling that continued—and was copied—for years. To create its first model, Keystone simply took a lamphouse from an 8mm projector and added a slide carrier and a lens. Few projectors had blower systems, and most of them warped slides.

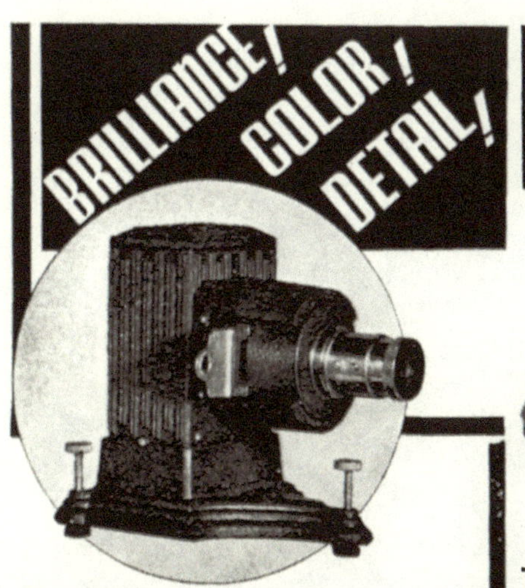

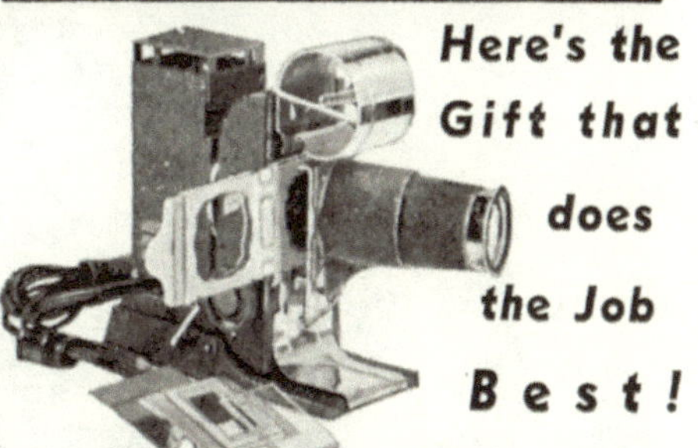

Advertisements featuring two popular projectors of the 1930's and 1940's.

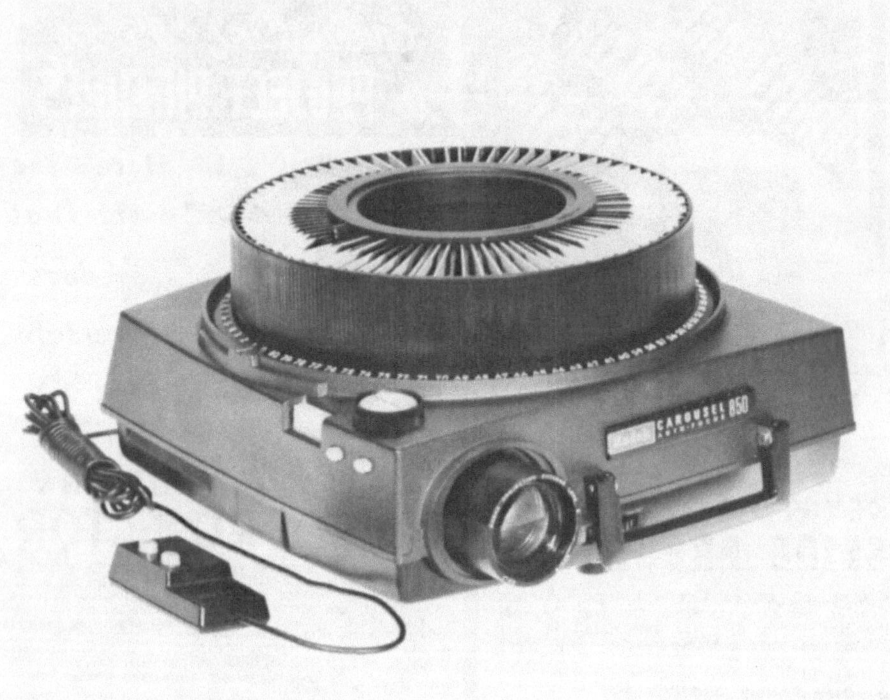

In contrast to the original Kodak projector, the Carousel 850 is low, sleek, and totally automatic. The development of this line of projectors was an engineering triumph for Kodak; the design virtually eliminated jamming and became the standard of the industry. But its relatively high price prevented the Carousel from becoming the volume leader. Sawyer's, Inc. (a division of the GAF Corporation), was the volume champion of the 1960's with a broad line of low-priced and private-brand models.

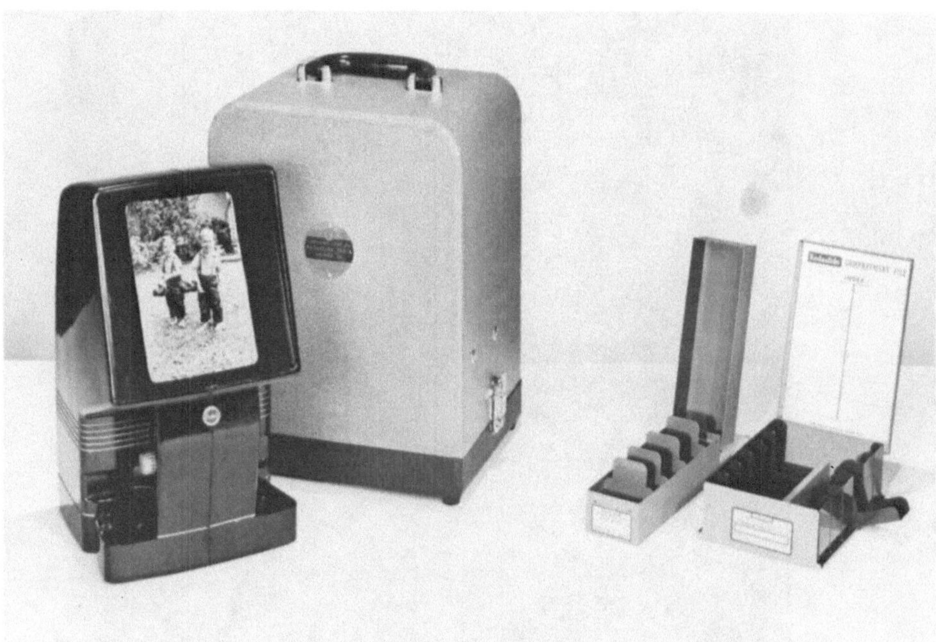

Left: Table viewers always seemed a good idea, but resolving the problem of rear projection—an optical requirement—resulted in a high price for the good ones. Almost all models were poor sellers. This Kodaslide Table Viewer, Model 4X, was typical of many. Right: A successful means of filing slides has never been found. Most people seemed to find the shipping container as good as any, and at one point Kodak actually returned slides in a plastic box that fit a slide projector. Slide files such as these are representative of the many available.

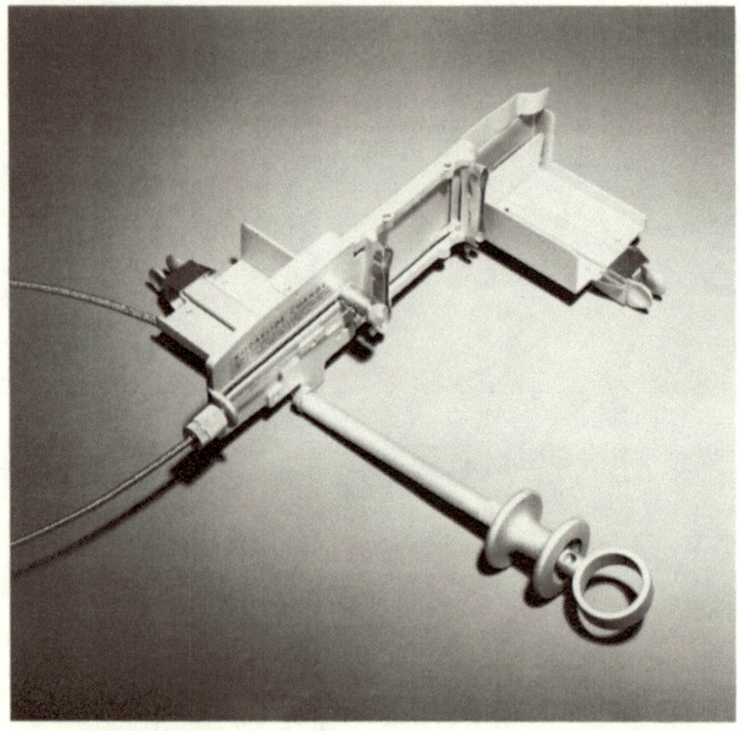

From the beginning, changing slides was an unhandy if not unpleasant chore. The Kodaslide Changer and several variations made by other manufacturers offered a convenient but not altogether successful solution, since, unless the slide mounts were almost perfect, the changers tended to jam.

(Projector evolution in the 1960's was an ad man's delight and the repairman's nightmare.) But after enjoying years of dominance in the color field, transparencies gradually gave way in public favor to the newer negative color films (Kodacolor appeared in 1942), from which both prints and transparencies can readily be made. No really satisfactory method of making good prints from color slides ever came along, despite the Kodak Minicolor (1941) and Ansco Printon (1945) processes. Today most prints made from slides go through an internegative process, by which they are first copied on color-negative material.

In 1944, Ansco brought out the first American-made color film which could be processed either at home or in a lab, and in 1947 Kodak Ektachrome film was introduced. But Kodachrome remained the favorite of most amateurs until it was replaced by Kodachrome II in 1961. Color film has traveled a long way from the introduction of Kodachrome in 1936, with a Weston rating of 8 (ASA 10), to today's Anscochrome 500, with a Weston rating of 400 (ASA 500), and it carried the American 35mm camera with it.

5 Flash Lit the Way

One of the 35mm camera's original appealing features was the ability of its fast lens to make pictures under available and adverse light conditions. Indoor picture taking at night soon became fashionable, however, and minicam fans were restricted to the illumination provided by photoflood lamps, using them in either cardboard or, if one was more affluent, metal reflectors.[1] The invention of the flashbulb in Germany in 1929 by J. Ostermeier for Philips (the Netherlands' equivalent of General Electric) held forth the promise of greater freedom for the 35mm fan, but it fell to professional photographers to pioneer and push the use of flashbulbs in the United States, even though most of them did not regard the canned light very seriously but as just another temporary fad.

Until August 1, 1930, when General Electric Company announced its No. 20 photoflash bulb, professionals and hardy amateurs used a magnesium dust called *flash powder*

[1] At 25 cents a pair the Kodak Handy Reflectors were used in bridge lamps. The reflectors were not altogether satisfactory and had a tendency to fall down on the hot bulbs, bringing dismay to the users—and, on occasion, fire departments. As the popularity of bridge lamps diminished, by 1939 so had sales of the paper reflectors. The Kodaflector, with its two aluminum folding reflectors, stand, and light sockets with cord at $5, was typical of the metal variety.

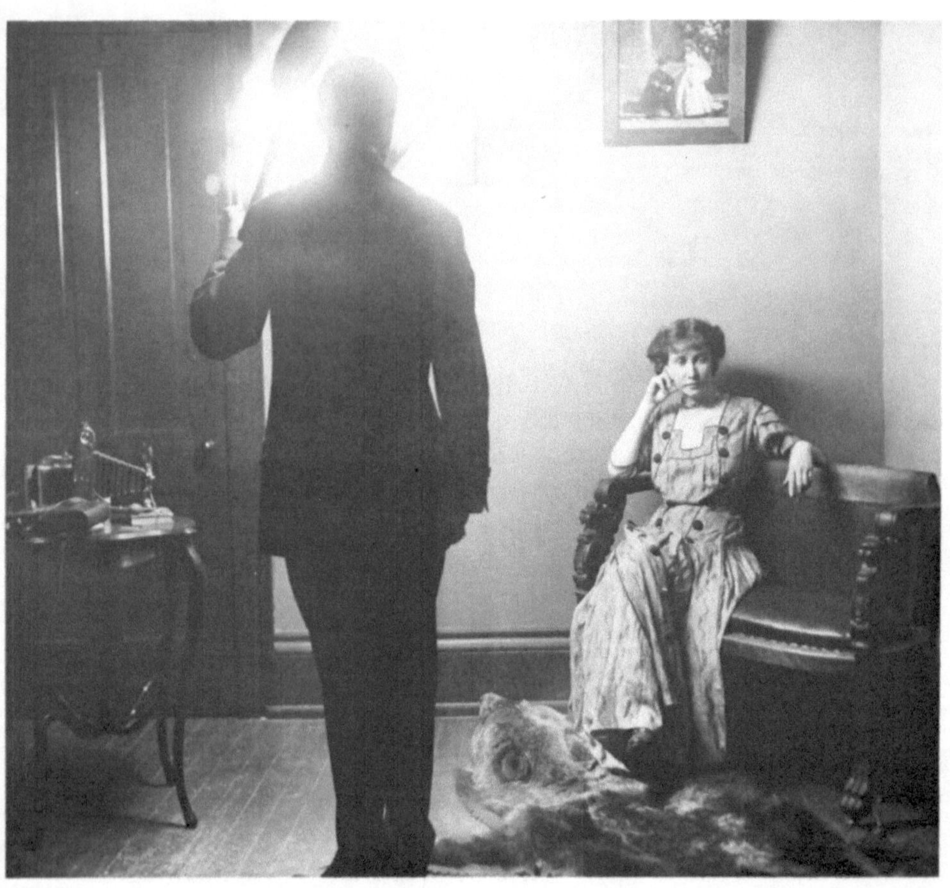

You would look the other way, too, if as much flash powder was used to take your picture as was apparently used in this publicity shot. Note the diffuser and the reflector. A more important objection, particularly to housewives, was the smoke which frequently left soot on the ceiling. In spite of these shortcomings, flash powder gave even illumination.

for illuminating outdoor pictures at night. Once the camera was positioned on a tripod and focused, the shutter was opened. The powder, spread in an open pan, was ignited, and the user then closed the shutter—a technique referred to as *open-flash*. Flash powder was erratic and difficult to control for a proper exposure. It was also dangerous. Accidents were numerous and often serious; one Hearst photographer lost an arm in an accidental explosion, and Hearst banned the further use of flash powder by members of his papers' staffs.

But then the flashbulb arrived. Filled with aluminum foil which ignited when a weak electrical current was passed through its primer, it gave a brilliant flash of illumination lasting but a fraction of a second. GE's first bulb was the size of today's 150-watt light bulb, and those willing to buy six at a time were given a one-third discount. Like flash powder, the early flashbulbs were used with the open-flash technique, and their main advantages were safety and ease of use. The early bulbs were inconsistent in timing, light duration, and output. These disadvantages were not noticeable when they were used under open-flash conditions, which gave great depth of field, but fast-action pictures remained out of the question. Press photographers gave flash photography a tremendous boost with various attempts to synchronize the flash of light with the shutter opening, an innovation which eliminated the need for a tripod and allowed hand-held exposures up to 1/200 second.

Men like "Doc" Skinner, of the *New York Journal*, Ernest Sisto, of the *New York Times*, and Clarence Stieglitz, of the *New York World Telegram*, had been experimenting with flashbulbs even before the GE No. 20 appeared. Sisto designed an electrical synchronizer unit which used six 1½-

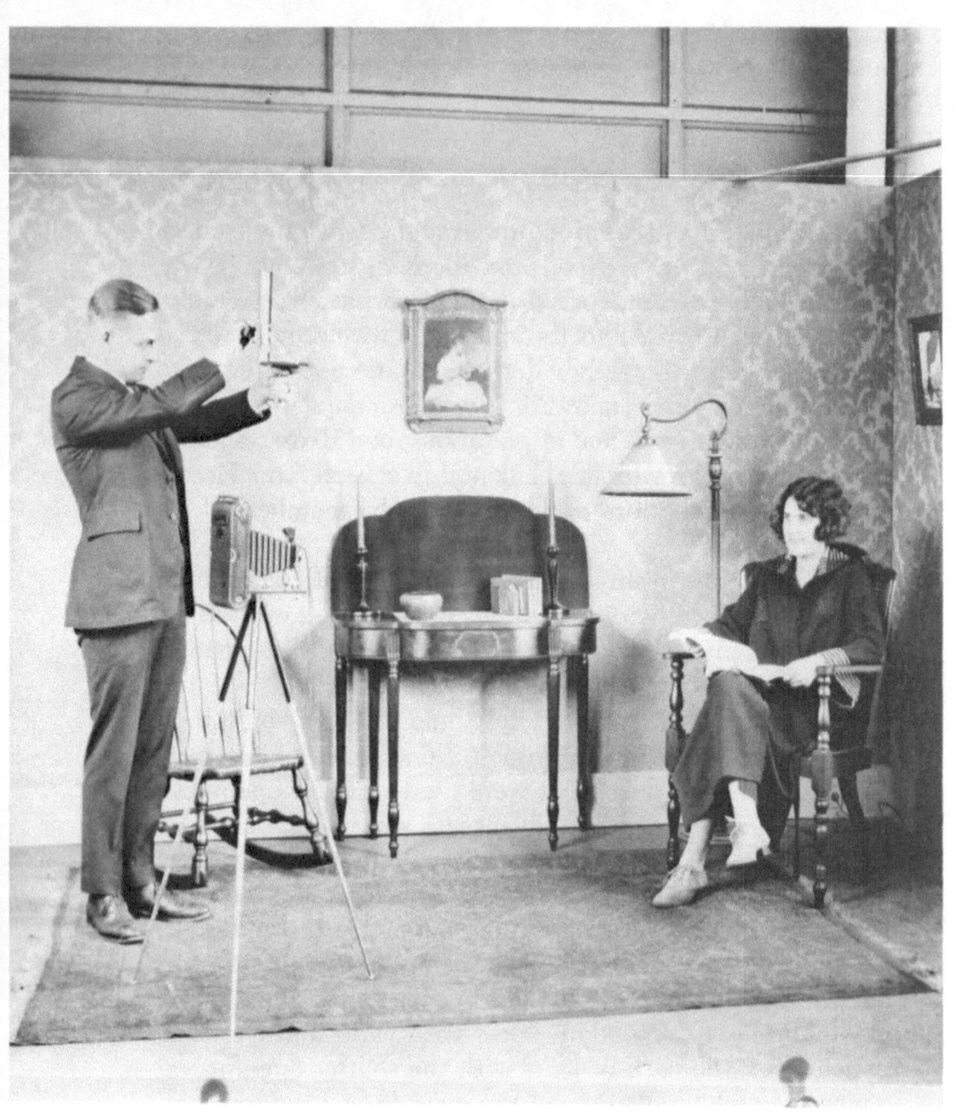

Flash sheets were a development beyond flash powder. Designed to make flash pictures easier and less dangerous, the sheets were held in a folding holder and ignited through a slot in the back. Flash tape was also used.

volt batteries to ensure a more nearly constant firing time. He sold about 50 of the units to other press photographers, and soon dozens of inventive photographers across the country were designing and selling their own synch units. Tommy Flannagen, of the *New York Daily Mirror*, produced probably the first "buzzer," or magnet, unit; Harry Hipwell designed his unit to incorporate a special air-compression release; and Hy and Morris Schwartz developed a mechanical synchronizer.

The first to move into full-time manufacturing, the Schwartz brothers and their father, Kalman Schwartz, owned and operated the Kalart Photography Studio in downtown Manhattan. In 1934 the family switched to synchronizer production and produced their Micromatic Speed Flash, a switch-contact unit built to fit into the cable-release socket of the camera.[2] In the following year Sam Mendelsohn, manufacturer of Bright Star batteries, opened a shop in New York and began selling his Speedgun, which immediately became the best-selling professional synch unit. Using a solid-core magnet solenoid mounted on the wooden lens board of the Speed Graphic, the Speedgun was mechanically linked to the shutter. Since the battery also tripped the shutter, it required more battery power to operate than the Kalart did but proved easier to synchronize properly and more reliable in operation. Finally came Sidney Lindahl, of Heiland Research, Inc., manufacturer of geophysical instruments. Lindahl's hollow-core solenoid gave the most accurate timing possible, and Heiland moved into the synchronizer business with a unit (available first

[2] The Micromatic permitted adjustments as fine as 1/1000 second; by turning the head of the Speed Flash, the user adjusted the tension of the spring behind the plunger until the shutter was released at just the right time to coincide with the flash.

on special order and in 1940 as a regular dealer item) which outsold even Mendelsohn's Speedgun and became the Cadillac of the flashgun industry.[3]

But the synchronizer only pointed up the flashbulb's many deficiencies. The firing time of the bulbs was not uniform, and each new batch had to be tested for *time lag*, that fraction of a second after the circuit was closed before the flash reached its peak of brilliance. A variation of 5/1000 second meant missing exposures if the shutter was set at 1/100 or 1/200 second. In an attempt to overcome this difficulty, Mendelsohn introduced a preheating synchronizer, which sent a low-voltage current through the bulb an instant before additional voltage was released to ignite the foil.

Bulb manufacturers were also busy making improvements. One of the early leaders in the field was Wabash Photolamp Corporation, the creation of A. M. Parker, a Brooklyn light-bulb manufacturer, and Phil Sperry, a newsreel cameraman. Their Wabash Superflash lamps, introduced in 1936, sidestepped the use of foil, utilizing instead a continuous length of fine wire made of hydronalium (a 7 per cent magnesium and 93 per cent aluminum alloy), precisely measured to assure constant light output. A big advantage of the wire-filled bulb over its foil-filled counterpart lay in the fact that the very nature of the foil made it impossible for the bulbs to give absolutely consistent light

[3] The era of the mechanical flashgun ended at the height of its technological advancement. The DeMornay-Budd Variable Focus Reflector was perhaps the most beautifully engineered flashgun ever built. Several attempts were made to perfect generator flash units to do away with the need for batteries, and both Kalart and Kodak introduced complete systems of flash equipment. The Kodak Ektalux flashgun was probably the most extensive, expensive, and ill-timed of them all; it was marketed too late, and few remember it today.

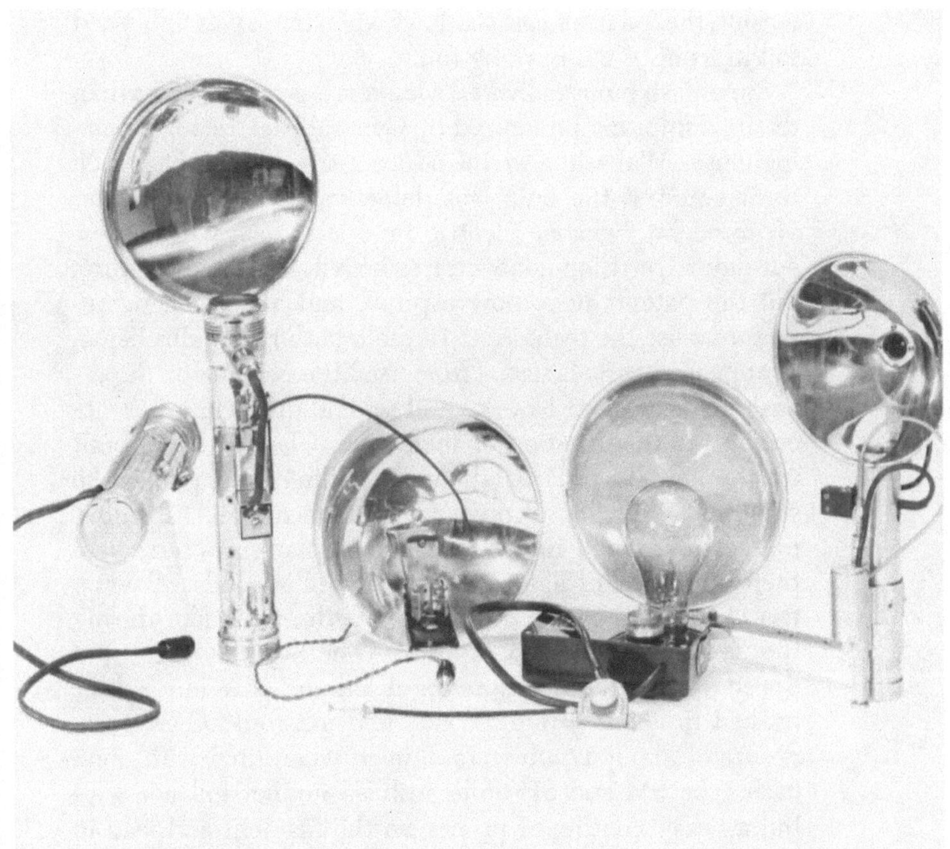

A pan attached to a flashlight was an acceptable early synchronizer. Some manufacturers even left the name of the flashlight on the bottom (note Eveready on the model at the left). The industry soon grew, and new ideas emerged. Two of the most popular synchronizers were the Kalart Micrometer (second from right) and the Mendelsohn Model E, which was also sold as the Kodak Senior Flash Gun (right).

output; the foil itself cast shadows when there was unburned foil in front of the burning foil.

Superflash bulbs delivered a longer, less intense light than the foil lamps manufactured by General Electric and Westinghouse. Wabash also introduced the blue spot which turned pink if the bulb was defective. Wabash was later acquired by Sylvania Electric Products, Inc., whose present-day advertising concentrates heavily upon this feature, but the patents have now expired, and almost all manufacturers use the technique. In the late thirties bulbs began to appear in various sizes (from the little No. 10 and Superflash O to the giant No. 75) and several different types, according to the duration of their flash. Class M bulbs built up the light to peak at about 20 milliseconds, giving the shutter blades time to open to their maximum. FP bulbs, marketed in 1939 for use with focal-plane shutters, held their high level of light for 30 to 60 milliseconds, allowing the focal-plane slit to travel the length of the film during the flash peak for even exposure. The Speed Midget and Speed Flash bulbs, whose light output was much less, peaked in about 5 milliseconds and gave an effective exposure of about 1/200 second, even when used with open flash. The SM and SF bulbs utilized neither foil nor wire but a heavy coating of primer on the filament and lead-in wires. Flash synchronization in most of the early inexpensive cameras was designed around these bulbs, which gave simple shutters the capability of stopping action. By 1940 the light output of flashbulbs ranged from 5,000 to 140,000 lumen seconds, and some were even available with a blue lacquer coating for use indoors with daylight color film.

The color temperature of flashbulbs is important when they are used with color films. That factor created a good market for Tiffen, Omag, Chess-United, Wratten, and other filter manufacturers. Most flashbulbs were rated at 3800° Kelvin, while daylight Kodachrome film was balanced for approximately 6000° and Type A Kodachrome for 3400° Kelvin, the color temperature of photoflood lamps. To maintain proper color balance, some filtration changes had to be made in the light used, and while blue-coated bulbs were standard from their introduction, the salmon-colored bulbs for use with Type A film caught on only with do-it-yourself fans, who also used dye solutions like Jen-Dip to coat their own bulbs.

For a short time during the infrared vogue of the thirties, flashbulb manufacturers marketed bulbs with a dye that allowed them to be used with infrared film at night or indoors. When they ignited, only a deep purplish-red glow could be seen, and then only if the subject happened to look directly at the bulb during the moment of flash. The fancy for the ghostly-appearing infrared portraits passed rather quickly, and the bulbs eventually became special-order items, with the dip manufacturers providing a special dye for home use.

As might be expected, entrepreneurs took advantage of this interest. Wonderlite bulbs appeared in 1938 in six different intensities, ranging from 75 watts (50 cents each) to 500 watts ($2.80 each). They were advertised as "providing a perfect 'north light,' the same quality of light that appears in the north sky on a clear day about 2:00 P.M., and may be used in total darkness or combined with daylight."

As flashbulb improvements were made during the thirties, prices began to drop, and by the close of the decade most bulbs could be purchased for 15 to 25 cents each. By that time several manufacturers were marketing synch units, and amateur cameras had started to appear with built-in flash. Users could choose from the flashguns marketed by James H. Smith and Sons Corporation (whose Victor brand was an established and respected name in the field of photographic lighting but whose heart remained outside the flash field), Goodspeed, Mendelsohn, Abbey, Jacobson, Schickerling, Heiland, Ulrich, Zenith, and Kalart, to mention only a few of the leading brands. While reflector shape, size, design, and finish were hotly debated topics among

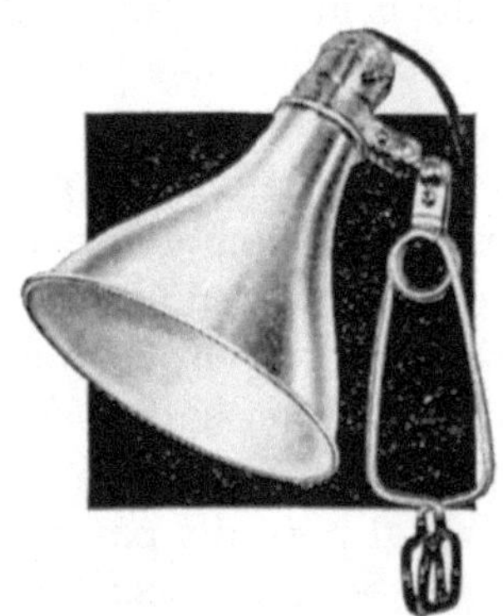

Newest and most effective of all miniature Fotoflood Lights

The

VICTOR MIDGET

Clamp-on Fotoflood Unit

ONLY

$1.50

Lamp Extra

Twice as effective as other smallest size lights. Has new shape reflector of polished aluminum and double ball swiveling mount with strong clamp and rubber covered grips that attach instantly to any convenient object. Uses No. 1 (25c) photoflood lamp.

Procure from your Camera Dealer or Write to 309 Lake Street for Folder.

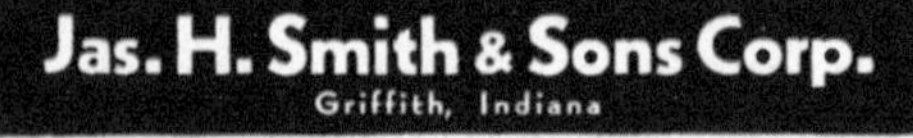

A typical clamp-on photoflood unit of the 1930's.

the professionals, the amateur market turned mainly to Kalart, whose prices began at $13.50. Although that was considerably below the price of most Heiland, Abbey, and Mendelsohn units, cost alone was not responsible for the popularity of Kalart synchronizers. The mechanical unit was easier to synch with the less-complex shutters and was certainly less cumbersome when attached to the smaller cameras. But the three companies soon outdistanced their other competition and, with Kalart, fairly well dominated the market through the late forties and into the early fifties.

Unlike the advent of Kodachrome film, whose effect was felt only in the view-camera, movie, and 35mm formats, the development of synchroflash cut across the entire spectrum of photography. It made possible inexpensive synchronized cameras and, combined with the introduction of color films which produced slides, brought many people into the ranks of 35mm camera owners. Flash also provided the instrument which, at least for a time, eclipsed the available-light school. Once professionals were convinced that flash photography was here to stay, they embraced it with a passion. "Flash" Casey became the stereotyped press photographer of the thirties, and no movie about a photographer portrayed him without his ever-present flashgun firmly attached to his Speed Graphic. And the 35mm minicam fans? Well, they split into two groups—those who embraced the new concept of lighting as an extension of the 35mm camera's versatility and the smaller number of die-hard adherents of available light, who grumbled that a high-speed lens was meant to be used.

After World War II flash photography was accepted by everyone, and by the fifties all new cameras came equipped with built-in flash synchronization. Gradually the electronic

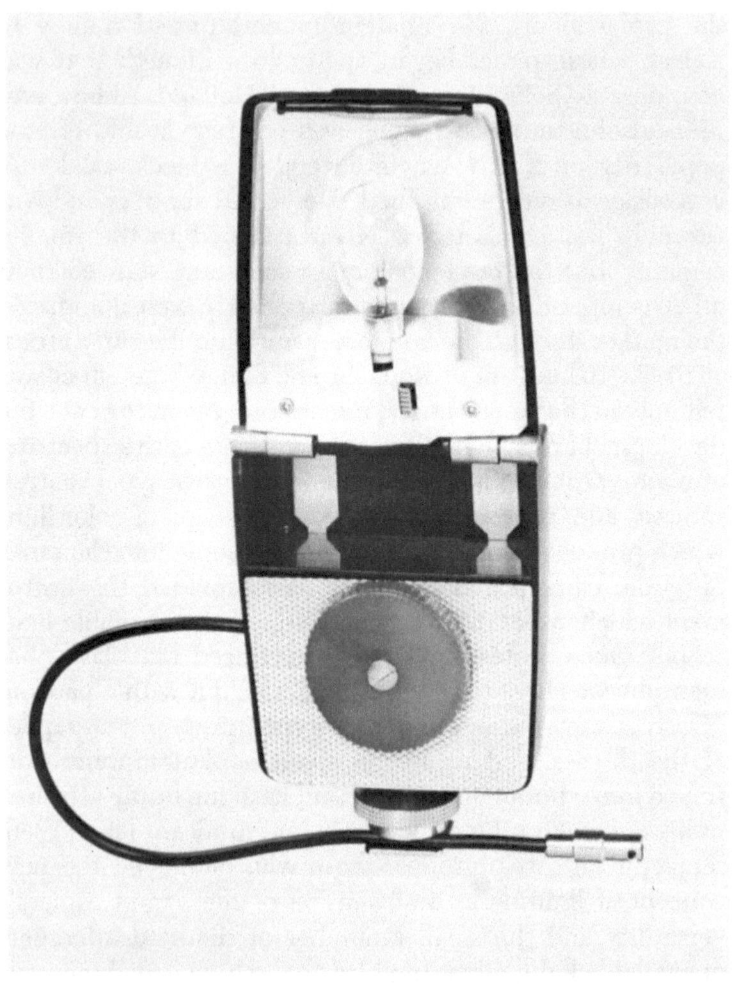

The Kodak Generator flashgun was an attempt to do away with batteries by means of a simple electrical storage unit. It was not the first attempt to incorporate a small hand generator in a flash unit, and like earlier efforts it met with consumer indifference.

flashtube (designed by Dr. Harold E. Edgarton, professor of electrical engineering in the Massachusetts Institute of Technology) replaced the flashbulb for the professional. Miniaturization of the bulb from the bayonet-base No. 5 and Press 25 to the M2 and the baseless AG-1 and now the flashcube and the magicube have increased its popularity among amateurs, even in the face of a growing flood of inexpensive imported electronic-flash units.

6 Let's Try *f*/8 at 1/50

Successful picture taking with any camera begins with correct exposure, and during the formative years of the 35mm camera a profusion of instruments appeared to help the photographer select the combination of lens opening and shutter speed which with proper development would produce the optimum negative. For many years exposure had been a hit-or-miss affair, beginning with a period of trial and error which eventually led to that wisdom we call experience. While the photographer developed a sharpened intuition, or "feel," for the correct exposure, it should be recalled that one of the major reasons for the development of the first Leica was the need for an exposure-testing device.

From the outset of photography individuals made use of prepared charts and tables, none of which told them very much (see Table 2). While manufacturers hedged on defining the exact emulsion speeds of their films,[1] various calculators were available to the bewildered photographer, usually in a tabular or slide-rule format which took into consideration such exotic factors as film type, lighting condi-

[1] Usually because film manufacturing at the time was more art than science and emulsions varied from batch to batch. By the early 1900's Kodak was able to recommend one exposure rather than an exposure range for its films when used under standard outdoor conditions.

tions, filters, time of day, season of the year, and latitude and longitude. Those using an exposure calculator would set it to take into account all the factors called for and then read its exposure recommendation from a scale. But by the onset of World War II it could honestly be said that film manufacturers and the makers of exposure devices had come close to establishing standards for each other.

The more sophisticated exposure calculators operated on a principle made somewhat popular for a time by the Watkins Bee, introduced in 1911 by Watkins Meter Company, of Hereford, England. This instrument came in a metal case the size and shape of a pocket watch and contained a small roll of sensitized paper. A small opening on the calculator face was bordered by a patch of gray paint, called a *standard tint.* The calculator was held in front of the subject to be photographed and pointed in the direction of illumination. A fresh area of the paper roll was moved into the opening

Table 2

Exposure Table, 1840's

Exposure Time, Seconds

LIGHTING	*1/6 plate*	*1/4 plate*	*1/2 plate*	*1/1 plate*	*24 × 32cm plate*
Sky with white clouds–facing north	2–4	10–15	15–20	20–30	30–120
Sky with white clouds–facing south	1–2	5–10	10–15	15–30	20–60
Open shade	1–2	5–12	10–20	20–40	60–90
Bright sunshine	Fraction of a second	1–4	3–6	6–10	15–20

beside the standard tint, and the user counted the seconds which elapsed until the paper color matched the tint. With this information he was able to set the calculator scale and read the recommended exposures. But the procedure was not quite as easy as it sounds. The paint patch was usually a bluish gray, and the paper turned a brownish gray, giving the photographer quite a bit of leeway.

A more sophisticated innovation was the optical meter, often referred to as the *extinction*, or *visual*, *meter*, whose origins are lost in the mists of time. Its scale contained a series of numbers or symbols (usually ten), each one a degree darker than the previous one. The scale was pointed toward the subject from the camera position, and the light reflected from the subject caused several of the numbers or symbols to become visible. Transposing the least visible one to the attached calculator, the photographer figured his exposure. Manufacturers of optical meters had to be very careful in arranging the meter scale; for example, if letters were used, they had to be out of alphabetical sequence to ensure that the least clearly visible letter would be used. Familiarity with the system often deceived the user into believing that he could read an additional number or letter, resulting in a grossly incorrect exposure.

A worse problem with which the user of an extinction meter had to cope was the factor of sensitivity. While the sensitivity of film to light remains constant, that of the human eye changes or adapts, a factor which also often led to an incorrect exposure calculation. Some such meters were held close to the eye like a telescope; others were viewed several inches from the eye. The scale, or *wedge*, as each section was called, proved to be another problem. Most wedges were photographic and not always the same from

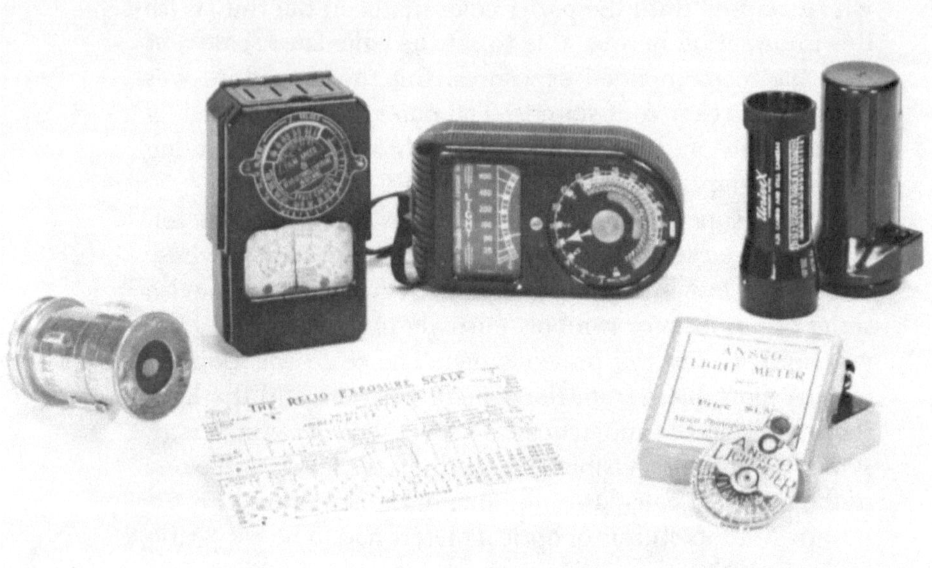

Determination of the amount of light necessary to expose film correctly was not helped by film manufacturers' reluctance to publish film speeds, but many companies came to the photographer's rescue with exposure meters and guides. Left to right: The Watkins meter used paper that was exposed to light and then compared with a known value. The original General Electric Photoelectric meter attempted a more scientific approach and, along with Weston, published a list of film ratings. Unfortunately, the two rating systems were not compatible. The UniveX was a traditional extinction meter with a graduated and numbered wedge, the last number visible being the exposure key (most people saw all the numbers). The Relio (below, left) was but one of dozens of scales and charts available. The Ansco was similar to the Watkins.

sample to sample; success with one camera did not necessarily mean that the photographer could use a friend's camera with the same results, even if they were both of the same make and model.

Optical meters were widely used, however, and many were attached to 35mm cameras (especially Argus and Universal models) for their advertising value or were made available as accessories which clipped onto the camera. Actually, while the optical meter could be very handy in many cases, it was really little more useful than an ordinary table or calculator. When the authors were younger, it was not uncommon to see photographers running all over the landscape pointing their optical meters after making an educated guess and never quite believing what the meters told them. While the extinction meters could be read with reasonable accuracy, they measured only the average brightness of light from the entire subject area and were incapable of determining the actinic quality of light, an important evaluation in terms of the film's color sensitivity. They also depended upon the human eye (a notoriously poor judge of light intensity) to determine the measurement by which the calculation was to be made. But they were inexpensive ($1.00 to $3.50), and their popularity lasted into the early fifties, adding a touch of low-priced glamour, if not accuracy, to the hobby.

Concurrent with the appearance of the Leica in the United States in 1932, the Weston Electrical Instrument Corporation introduced the first photoelectric light meter. It gave accurate electromechanical measurements of the amount of light present, and no optical juggling or mental computation on the part of the user was necessary. The meter functioned with a selenium photocell which converted the light reaching its surface into a tiny electrical current and moved a galvanometer needle across a calibrated scale; this reading was then transposed to a calculator on the meter, from which exposure data could be read directly.

The color sensitivity of the photocell was closely matched

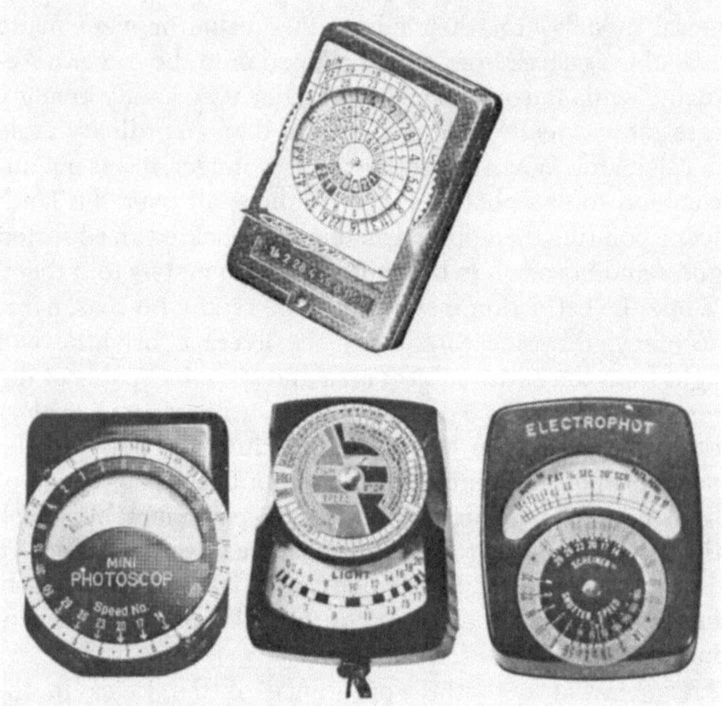

Above: The Expophot extinction meter was designed for use at chest level and cost but $1.75. Below, left to right: The Mini Photoscop ($14.75), the DeJur Amsco ($10.50), and the Electrophot ($10.00) were photoelectric meters noted for their convenience and low price, but they offered the Weston line little sales competition.

to that of the film, and its angle of view was similar to that of a camera lens. These innovations overcame two major faults of the optical meter. While almost any photoelectric meter would give an acceptable performance in bright light, meter sensitivity was closely related to price, and the more costly the meter the dimmer the light conditions under which it would operate properly. Some of the more expensive meters had accessory amplified cells to increase their sensitivity under poor lighting conditions.

The photoelectric light meter was not really a new discovery. Its principle was known in England as early as 1875, and in 1881 the Englishman C. J. Burton attempted to use a selenium cell for tripping the shutter and adjusting the exposure at the same time, a feat not accomplished satisfactorily until 1938 with the introduction of the Super Kodak Six-20 camera. The concept of completely automatic exposure introduced by the Super Kodak Six-20 was sensational for its time. The idea was sound, but the technical problems, particularly in miniaturization and electronic circuitry, made operation erratic. Alan Stimson went to Kodak from General Electric and worked on the problem of the minimum metering requirements for such a camera. His statistical calculations, based on thousands of "typical" snapshots, are still the basis for automatic cameras.

Within a few years after the introduction of the photoelectric light meter at least fifteen different American and foreign models were available. The DeJur–Amsco, Photrix, and Electrophot were popular, but the Weston line was far and away the most popular and was seriously challenged only briefly by the introduction of the General Electric meter in 1937.

Meter manufacturers required some method of classify-

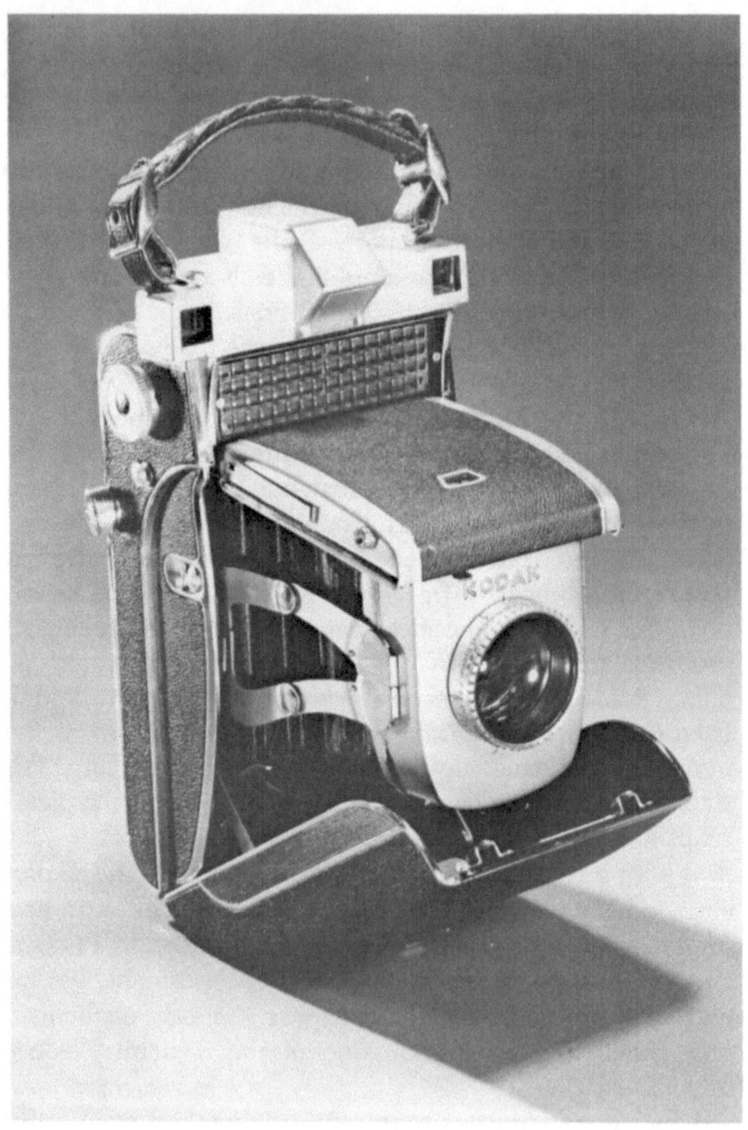
KODAK

ing film speed for the basis of exposure computation, and in this country at first only Weston undertook to provide a rating for each film. That was one of the major reasons for the popularity of Weston meters; meters manufactured by other companies in the United States employed the Weston system, in which doubling the rating meant doubling the film speed. Thus a film rated at Weston 64 was twice as fast as a film rated at Weston 32, a difference of one lens stop.

The many foreign meters being marketed in the United States introduced three more rating systems with which the user had to contend—DIN (for Deutsche Industrie Normen, or German Industrial Standards), degrees Scheiner, and Hurter and Driffield (H and D). To make matters even more confusing, there were two versions of the Scheiner system, the European and the American. The European rating was two full lens openings higher, and a film with a European speed of 24° Scheiner was really only 21° American Scheiner. This variation caused much confusion among American tourists abroad using an American or export-model meter

While not a 35mm camera, the Super Kodak Six-20 was the first automatic camera and established the basic system for all others to come. Introduced in 1938 as a part of a series of fine cameras (which included the Kodak Bantam Special), it was beautifully engineered, but its $225 price tag put it far above the average man's reach. Camera fans today are amazed by the features and the quality of workmanship, all of which seems to prove Kodak's long-held contention that American manufacturers could produce anything photographic that the public wanted and was willing to pay for, including superior construction and innovation. Perhaps the most amazing thing about this camera was that Kodak was willing to manufacture it, knowing its technical limitations and the limited market potential.

and purchasing their film along the way. Since each film-speed-rating system derived its values in completely different ways, no accurate mathematical conversions were possible. Translating a DIN or an H and D rating into a Weston figure was only an approximation based upon information such as that shown in Table 3.

Table 3

Film-Speed Rating Systems, About 1938
Approximate Conversions and Comparisons

WESTON	*DEGREES SCHEINER*	*DIN*	*H and D*
4	15	8/10	190
5	16	9/10	240
6	17	10/10	308
8	18	11/10	390
10	19	12/10	500
12	20	13/10	636
16	21	14/10	800
20	22	15/10	1,050
24	23	16/10	1,300
32	24	17/10	1,700
40	25	18/10	2,100
50	26	19/10	2,700
64	27	20/10	3,500
80	28	21/10	4,400
100	29	22/10	5,600
128	30	23/10	7,200

The matter became even more confused when General Electric announced its own rating system in 1939. While Weston had apparently adopted the Burroughs-Welcome exposure system, which used fractional values (1/32) and made them into whole numbers (32), GE seems to have simply added one-third more to the Weston numbers, making a Weston rating of 8 appear as 12 on the GE scale. And if all this confusion was not enough to drive the poor amateur to the brink, Weston periodically revised its ratings and published new tables, with which Kodak and other film manufacturers chose to disagree at the time. Even developer manufacturers produced their own emulsion ratings, as illustrated in Table 4.

Until the advent of the rating system devised by the American Standards Association, which most film, bulb, developer, and meter manufacturers adopted after World War II, it was every man for himself and anyone's guess. The photographer was forced to make tests with his own equipment to determine exactly which rating provided the best negatives or slides with his favorite combinations of film, bulb, and developer.

Table 4

Emulsion Ratings for Selected Films, Calculated in 1938 by the Weston System

Film	*Manufacturer's Recommendation*	*When Developed in D-76*	*When Developed in Champlin 15*
Agfa Supreme	80	40	100
Agfa Ultra Speed	128	64	144
Dupont Superior	64	24	64
Gevaert Panchromosa	32	24	40
Kodak SS Pan	32	24	40
Kodak Panatomic	13	16	20
Perutz Peromnia	32	24	40

Source: Compiled from published charts of various manufacturers of films and developers.

7 Fine-Grain and Frame It

Along with research in film emulsion and speed during the thirties, developers and the developing process also underwent much experimentation in both research laboratories and home darkrooms. A detailed discussion of film developing with its many variations and ramifications is beyond the scope of this book. We shall merely touch briefly on the theoretical to place the historical in proper perspective. The purists among our readers may well point out that we are discussing the chemical and physical change which occurs in development as if one or the other were absolute. So far as we know today, this is not possible, but for the sake of simplicity and to avoid further confusion we shall consider the dominant action.

There are two principal methods of film developing: *chemical development and physical development.* Chemical development is used almost altogether today. The method makes use of a *developer*, a solution containing a number of dissolved chemicals which act upon the exposed film and "bring out," or make visible, the photographic image. A film emulsion contains many crystals of silver halide (primarily silver bromide) suspended in gelatin. The sizes of the crystals vary from one kind of film to another and within a single emulsion. To ensure satisfactory tone reproduction, the crystals in a standard film vary from about a fifteen-to-one ratio in size distributed evenly throughout the emulsion. The larger crystals are more sensitive to light, which means that in faster films the average crystal size is larger and that during exposure they are exposed first.

While the film emulsion is composed of silver halide crystals, the photographic film image is formed of bits of metallic silver scattered through the film emulsion in a manner determined by the location of the crystals, the exposure, and the development of the film. These bits of silver are produced by the action of the developer on the exposed crystals, which gradually changes them into black metallic silver. Several factors determine the average size of the bits of silver: the size of the original crystals, the amount of exposure, the "degree" of development, the temperature during processing, and the type of developer.

A small camera simplifies picture taking, but people are not satisfied with the little *contact* (same-size) prints. Small negative size makes enlarged prints necessary. When the photographic image is enlarged enough, the basic structure of the negative photographic image—composed as it is of discrete bits of metallic silver—becomes apparent in the print. Rarely is the degree of enlargement great enough to show the individual bits of silver scattered throughout the film. What does show is the *grain structure* caused by the location of overlapping pieces of silver, sometimes called *clumps of silver.* A typical film emulsion may average ten crystals thick when manufactured. Some tiny areas of film have no crystals; other areas have several from top to bottom. Where several crystals overlap, several specks of silver will be formed in development, and a grain will result.

When early-day films were enlarged only a few diameters, the grain structure was apparent in the print. Film manufacturers began making emulsions with smaller crystals, which slowed the film but made the structure finer and reduced the sizes of the clumps of silver. Crystals do not change position during processing, as the term *clumping* implies. The clumps are predetermined in the manufacture of the emulsion. The size of the bits of silver can also be affected by the type of development the film receives, and the smaller the bits the finer the grain.

The chemicals usually included in a film developer are as follows:

1. *Developing agents*, sometimes called *reducing agents*, which change the exposed silver halide to black metallic silver. The two most common developing agents, Metol and hydroquinone, are normally used together. Metol is high-energy and quick-acting but allows a minimum total degree of development. Hydroquinone is slower-acting and develops a greater degree of contrast and a greater maximum density.

2. *Activator*, an alkali which makes the developing agents work better and faster. Developing agents work best in an alkaline solution; the greater the alkalinity the faster the development rate. Activators are chemical "bases" such as sodium carbonate, borax, and even caustic potash (potassium hydroxide).

3. *Preservative*, usually sodium sulfite, which slows oxidation of the developing agents.

4. *Restrainer*, usually potassium bromide, which tends to prevent the development of unexposed grains of silver halide, which would cause fogging of the film.

By experimentation it was found that the use of very mild

alkalis reduced the grain size when the hydroquinone content was also reduced. It was also found important to work for thin, relatively low contrast negatives. This was achieved by giving full exposure and reducing the degree of development. Another successful approach was the addition of a silver halide solvent to the developing solution, usually potassium thiocyanate. This chemical began dissolving the crystals as they were developing, limiting the size of the silver bits, as well as the maximum density attainable.

It was also discovered that paraphenylenediamine, another developing agent, tended to reduce the grain size. This isomer makes a "low-activity" developer and acts as a silver halide solvent as well, both of which reduce grain size. It also lowers the film speed considerably when compared with the Metol-hydroquinone formula, usually by two stops or more.

In the thirties various "fine-grain" developers, mostly secret-formula proprietary products, appeared. They successfully reduced the grain size so that enlargements of reasonable size could be made from 35mm negatives. But the really important advances in fine grain came from breakthroughs in emulsion manufacture. The original method of gaining speed in an emulsion was to make the crystal size larger. Methods of chemically sensitizing the crystals by the addition of sulfur or gold compounds permitted increased speeds with small crystals, a real boon to the 35mm fan. Practical film speeds with finer grain—or greater film speeds with the same grain size—became a reality in the thirties with the introduction of the "atomic films" (a term referring to the very small size of the silver bits).

In the fifties another breakthrough occurred in the manufacture of fine-grain films. The thin-emulsion films which increased sharpness to a great degree also showed a further decrease in grain size. With these films, carefully processed, enlargements twenty times the size of the negative became possible with grain held to a very tolerable level—20 × 30-inch prints from a 1 × 1½-inch negative. In the meantime, film speeds were increased tremendously while the grain size was held steady. Today's standard high-speed film (ASA 400) has finer grain than the first atomic film and over ten times the speed. The fast films with reasonable fine grain greatly outsell the slower films with the superfine grain.

By 1938 the darkroom bug was overwhelmed with pre-packaged fine-grain developers.[1] All of them had merit for the functions they were called upon to perform. The user could choose Micrograin "85," Edwal 12, Edwal 20, Imperial, Gamma "D," Minicol, X-33, Marshall's, D&F, G.D.X., Perutz UFG, Defender Panthermic 777, Champlin 15 and 16, Kodak DK-20, and dozens more.[2] Many of the "super" developers relied upon the simple expedient of overdevelopment, a process accomplished as easily with underexposure and a standard developer. Others worked on the basis of overexposure and underdevelopment, reducing contrast in the process.

Development soon became a matter of highly individual preference, and many heated arguments took place between amateur photographers, in the pages of their magazines,

[1] He was also swamped with treatises published in book and pamphlet form which expounded the virtues of certain solutions within the context of a general consideration of the theory of development. Of these expositions *Champlin on Developing*, by Harry Champlin (1937), was one of the best.

[2] Gamma "D," developed by pictorialist William Mortensen and Dr. Albert Doran, was advertised as the "atomic" fine-grain developer. American Scientific Products' D&F was a combination developer and fixer, or monobath.

and on speakers' podiums at camera-club meetings. While most darkroom users adopted the standard prepared formulas, many others preferred to formulate their own, and research into the profusion of individual and "secret" formulas used at the time leaves one properly dizzy today.

Three of the most popular prepared developers were Defender's Panthermic 777, Champlin 16, and Kodak DK-20, each having its own band of boosters and each possessing a claim to fame. Panthermic 777 could be used at temperatures from 60° to 90°F, with satisfactory results up and down the range. Each time a developing solution was used, it lost some of its strength, and replenishment was necessary to restore its strength. Replenishment—the addition of a quantity of developer after each use—was an inexpensive and popular way of prolonging the life of a developer. Unlike the replenishers of the other formulas, 777's replenisher was not the base developer in a concentrated form but contained a different chemical balance, which seemed to make sense to those of us looking for something better than what we had been using and bewildered by all the claims and counterclaims about the "best" developer.

Harry Champlin's Formula 16 contained only three chemicals, sodium sulphite, chlorhydroquinone (acid-free), and tironamin "C," and could be used between 64° and 81°F, an improvement over his Formula 15, which restricted users to a 4° range. It was a concentrated developer mixed nine-to-one with water, used once, and then discarded.

Kodak's DK-20 developer introduced the use of sulfocyanate and a new chemical especially compounded for it, Kodalk, which replaced the sodium carbonate commonly used in developing formulas. Kodak DK-20 had the added virtues of excellent keeping qualities, much finer grain than the popular Kodak D-76, and a minimum loss of emulsion speed (only 35 per cent when compared to D-76).

Amateurs who compounded their own formulas experimented with such chemicals as paraphenylenediamine, a poisonous staining chemical which had excellent fine-grain qualities but whose instability made it highly unpredictable in use; glycin, also an unstable substance, often combined with paraphenylenediamine to take advantage of its preservative action and used mainly to reduce emulsion-speed loss in development; pyro (pyrogallol), one of the oldest reducing agents known to photographic chemistry; pyrocatechin, a close relative of pyro whose tendency to soften the emulsion greatly increased the chances of inadvertent reticulation; and chlorhydroquinone, a variation of hydroquinone which, unlike the latter, did not cause severe clumping of the silver. With all of the frantic activity of the thirties in the field of negative development, it is interesting to note that careful photographers could obtain nearly identical results with any fine-grain solution, regardless of the combinations of film and developer by which they swore.

An old process of film developing, called *physical development*, was updated by Dr. Allen F. Odell and enjoyed a brief vogue in the late thirties. Physical development, a method of silver-plating the negative's latent image, resulted in a tonal-gradation scale proportional to the contrast in the original subject and required no increase in exposure. It differed from chemical development in that the silver used to form the image was derived from the solution itself. A prebath of potassium iodide and sodium sulfite was used before the developer. The developer itself, called the *silver-stock solution*, contained sodium sulfite, silver nitrate crystals, fresh sodium thiosulfate (hypo) crystals, and water, diluted to a working solution no more than ten minutes before use by the addition of more water and amidol.

While physical development appeared to produce finer-grain negatives than the chemical process, it was a mysterious and rather difficult procedure, demanding particular attention to the mixing of the stock solution, filtering of the developer before use, and the complete removal of the silver residue which accumulated during development. In addition, time and temperature requirements were strict; the process required about one and a half hours to complete with a maximum temperature variation of only 1½°F. Those who accepted the limitations produced exceptionally fine grain negatives; uniform in quality, they could be enlarged greatly without loss of *acutance* (image sharpness). But most darkroom workers chose the less exacting and time-consuming chemical development. After all, dedication is dedication, but who wanted to spend his whole life in a darkroom?

Ah, the darkroom! Mention of it brings back fond memories to those of us who discovered the magic delights of photography under the red safelight. This indoor ritual was performed for about a decade and a half before slowly declining in favor, primarily as the result of an affluent economy, jumbo prints from processing laboratories, and, most of all, the advent of negative color films and color prints.[3] Some people attribute the decline of the darkroom hobby to the advent of television in the late forties, and others believe that rising costs of equipment and supplies helped de-

[3] A slow but steady resurgence in darkroom interest has been noted in recent years as manufacturers continue to simplify the processes for making color and black-and-white prints at home.

liver the *coup de grâce*. Whatever the reason, people found other ways of spending their evenings.

In the early days of 35mm, no minicam owner could really fancy himself a photographer unless he did his own developing and printing; it became something of a status symbol within the ranks. Although it was a practical and sometimes financial impossibility for many, the true mark of the 35mm devotee was his own darkroom, whether a specially designed room in his house, his wife's kitchen, or a spare closet.

For those who did not have their own equipment or were not challenged by the creative possibilities of the darkroom, commercial laboratories appeared across the nation to lend a helping hand. The mass-market photofinishing industry provided a new and large area of endeavor for photographic-equipment manufacturers like Pako, of Minneapolis, and Eastman Kodak Company. Those companies transferred the engineering skills they had developed in the design of continuous processing and printing machines for the motion-picture industry to the creation of a whole new line of semiautomated developing and printing equipment.[4] That it became a highly profitable market could not be denied. In 1935, Leo Pavelle, who was to become an outstanding leader in amateur and commercial color finishing a decade later, had a small laboratory in New York which produced an average of 5,000 minicam prints a day. Minilabs, Universal, Seaboard, and other labs across the country were kept busy providing enlargements from 35mm negatives at no greater cost than that for contact prints from larger negatives.

Competition among labs was keen, and the cost-conscious

[4] Kodak's Kodaprinter made 700 jumbo prints an hour from the small negatives at a retail cost of 3 to 5 cents each.

consumer had his choice of them. For $1 (wrapped around the film and held with a rubber band in those days before we became a nation of check writers[5]) the mail-order labs would develop a 36-exposure roll in a fine-grain solution and enlarge each negative on single-weight glossy paper to 2½ × 3½ inches, 3 × 4 inches, or 3½ × 5 inches, depending on the lab. The price was constant among labs, but with a bit of shopping around it was not too difficult to find one that made slightly larger prints or perhaps used double-weight matte paper. An extra 50 cents brought a reloaded cartridge of film (usually Afga Superpan or Du Pont) in the return mail with the prints.

But for those of us who rose to the challenge of the darkroom, an entirely new aspect of photography lay before us. Here was the one opportunity to glorify our hobby, producing trophies to hang on the wall, exhibit in salons, enter in competitions, and exchange for cash if the contest judges found our talents worth rewarding. It was our chance to correct mistakes we made when shooting our pictures, to rearrange composition, and to salvage errors in exposure. Under the glow of the safelight the magical appearance of an image recorded only hours before gave the photographer a sense of creation which must have been akin to motherhood. It was almost hypnotic in its effect; there was always one more exposure to try and just one more print that would be perfect. Many a time the radio station played the National Anthem and went off the air—an unheeded warning that bedtime had long since passed. It was difficult to schedule an evening's work.

In addition to film development and negative enlarge-

[5] Even today processing labs receive payment for as much as 75 per cent of their orders in cash—a constant source of temptation for mail thieves.

ment, which allowed him to experiment with a wide variety of chemicals and paper surfaces, the darkroom enthusiast spent many long hours—and long nights—working with reducers, intensifiers, and desensitizers. These were old processes long known to photography, but the tendency of minicam users to push films to their practical limits and beyond brought them a new popularity.

Intensification was especially favored, for it could be used to salvage an underexposed or underdeveloped negative by building up its density. But J. H. Smith & Sons, who marketed Victor Intensifier, also recommended a *deliberate* underdevelopment in their own formula (a 3-minute solution), a thorough fixing and wash, and then 15 to 60 seconds of intensification as a means of building contrast without grain. Reduction, the opposite of intensification, was not quite as popular, for overexposing was not as common as underexposing, and the length of time required for fine-grain developing helped guard against inadvertent overdevelopment. If one lost track of the developing time, his tendency was to remove the film, usually before development had in fact been completed—creating a need for intensification.

Few orthochromatic emulsions (which could be processed under a red light) were available in 35mm, and panchromatic film required development in total darkness or in a very weak green light. Since most of the available developers had a rather high fog index, those who developed their negatives by inspection (tray development under an amber or red safelight) favored the use of a preliminary desensitizing bath to eliminate the possibility of fogging during development. After giving the film a two-minute prebath in total darkness in Desensit, Agfa Green, Marshall's, Burrows-Welcome, or a desensitizing agent concocted by the darkroom operator, he developed the film. These solutions also removed the backings which film manufacturers had begun using to prevent *halation*, a blurred effect at the edges of a light area on the film caused by the reflection of light from the surface of the film. The backings would have interfered with inspection. Some of the desensitizing baths

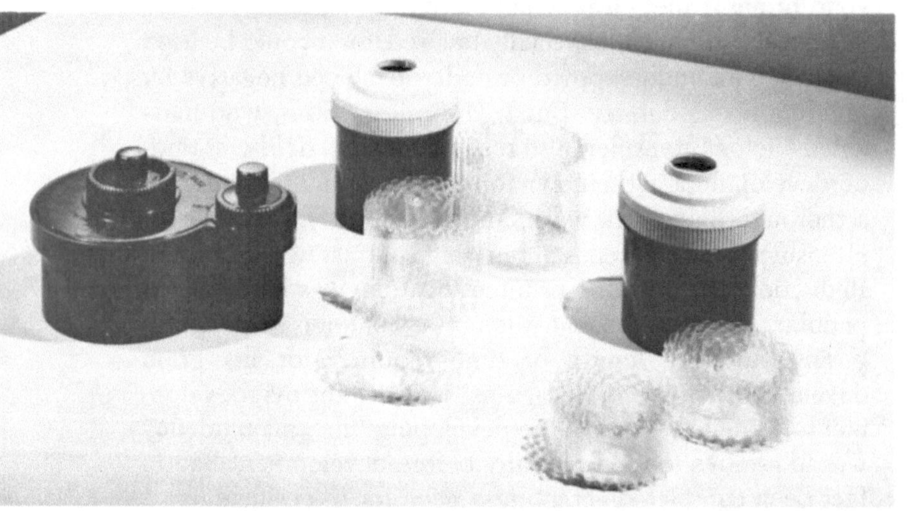

With the introduction of roll film a new means of easy film development became necessary. One of the first was the use of aprons around which the film was rolled before it was submerged in the developer. Early aprons had rubber tracks along the edges to suspend the film; in the modern versions shown in the illustration, the aprons merely have indentations. The invention of tank reels eliminated the need for aprons but made threading more difficult until an accessory was devised to lead the film through the grooves. The Kodak Daylight 35mm tank above accomplished this by attaching the film to the center of the reel and loading it from the inside out.

made it possible to develop the film in a very dim and shaded white light.

Even though they were accustomed to tray development of roll films, most amateurs found the 70-inch length of 35mm film too unwieldy to process in this manner, and so the 35mm developing tank made its appearance. Roll-film tanks were invented almost as early as roll film itself, and it was easy to devise or adapt a tank to accommodate 35mm film. Many of the darkroom tools in use today have not evolved much beyond their original form. The developing tank falls into this category. It is a round container with a locking cover which uses a grooved reel (or a plastic apron with crinkled edges) to keep the film from touching other surfaces, allowing solutions to flow freely across all film surfaces. Once the film is threaded onto the reel and deposited in the tank and the cover is locked, the room lights can be turned on and the darkroom operator can complete the developing process under normal illumination. Solutions are changed through a baffled opening in the cover of the tank. Daylight-loading developing tanks made a brief appearance, but many had the habit of not working correctly, and serious amateurs did not pay much attention to them. While expensive stainless-steel units made by Nikor, of West Springfield, Massachusetts, and Kodak were available, most manufacturers made use of the new plastics. For those who did not mind stretching their arms, Kemp Camera Supply brought forth its Devel-O-Tray, a small plastic container shaped like half a log on legs with a hold-down roller to keep the film from slipping out of the solution (only 4 ounces) as it was seesawed up and down.

The area in which most innovation occurred was that of enlarging. Before the advent of 35mm, enlarged prints had

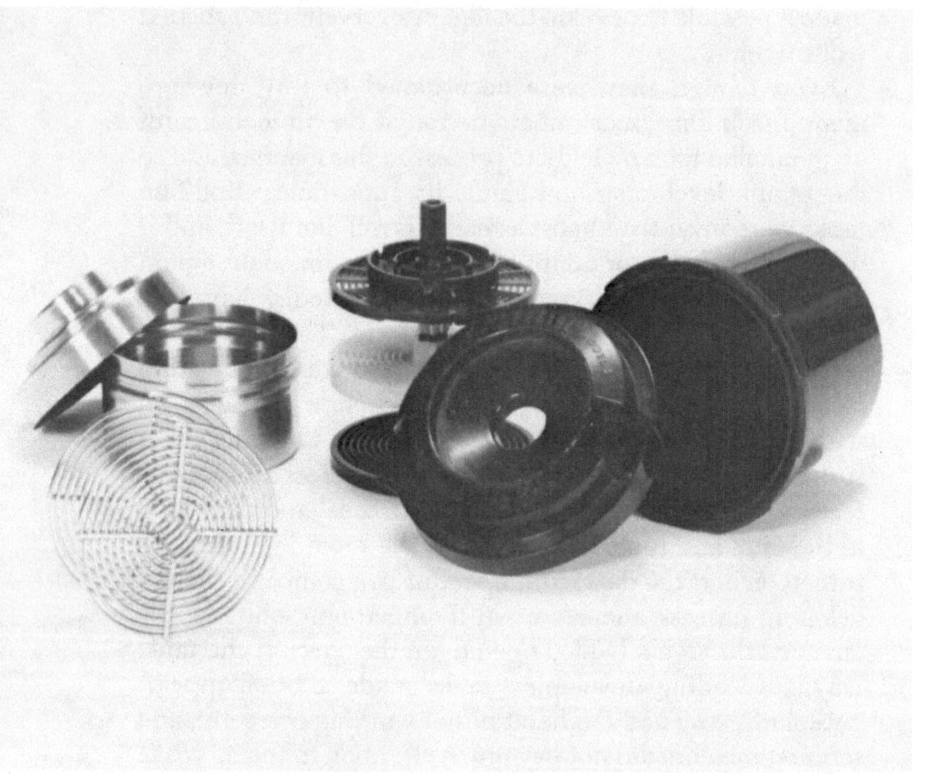

Difficult for beginners to thread, the Nikor stainless-steel tank (left) was and still is popular with professionals. While the name sounds like a Japanese import, the Nikor has always been an American product, although many imported copies have arrived since the patents expired. Perhaps the most popular tank ever manufactured is the F-R Special (right), manufactured by the F-R Company, Inc., of New York. Made of plastic but guaranteed for life, the accessory 35mm reel allows two rolls to be developed at one time. The careful photographer can load two sets of films back to back and develop four rolls simultaneously.

not really been necessary—most negative sizes gave a satisfactory contact print for viewing purposes. Early enlargers, bulky in design and clumsy in use, were used mostly by professionals (and most professionals really preferred to contact-print their work). But the virtual impossibility of satisfactorily viewing a 35mm contact print soon made the enlarger an amateur tool as well, with manufacturers designing and producing a wide variety for the home darkroom.

Among the most popular from a sales point of view were the Kodak Precision Enlargers, made in sizes to accommodate 2¼ × 3¼-inch and 4 × 5-inch negatives. Later a portable apartment-size model for 35mm which could be taken apart for storage was made available. Sturdy, dependable, and efficient, the Precision line was well regarded in spite of its bulging lamphouse.[6] But the name usually regarded as the Cadillac of the decade was the Omega, introduced by Simmon Brothers in 1938. With its aluminum lamphouse, low-voltage bulb, double condensers, glassless negative carrier, and counterbalance for easy raising and lowering, the Omega A was a thoroughly professional darkroom tool in miniature at a reasonable price ($48 plus lens). Its appearance and functional design have been retained to a surprising degree in current Omega models, giving it a sort of timelessness that can only be equaled by the Solar enlargers of Burke & James, which seem to have arrived with the advent of photography and are today much the same in appearance as their predecessors.

[6] Users of this particular enlarger were a hardy breed. When the operator leaned forward to check focus and composition on the easel below, the odds were great that unless he was wary his forehead would come to rest against the hot lamphouse, resulting in a mighty scare, if not a burn.

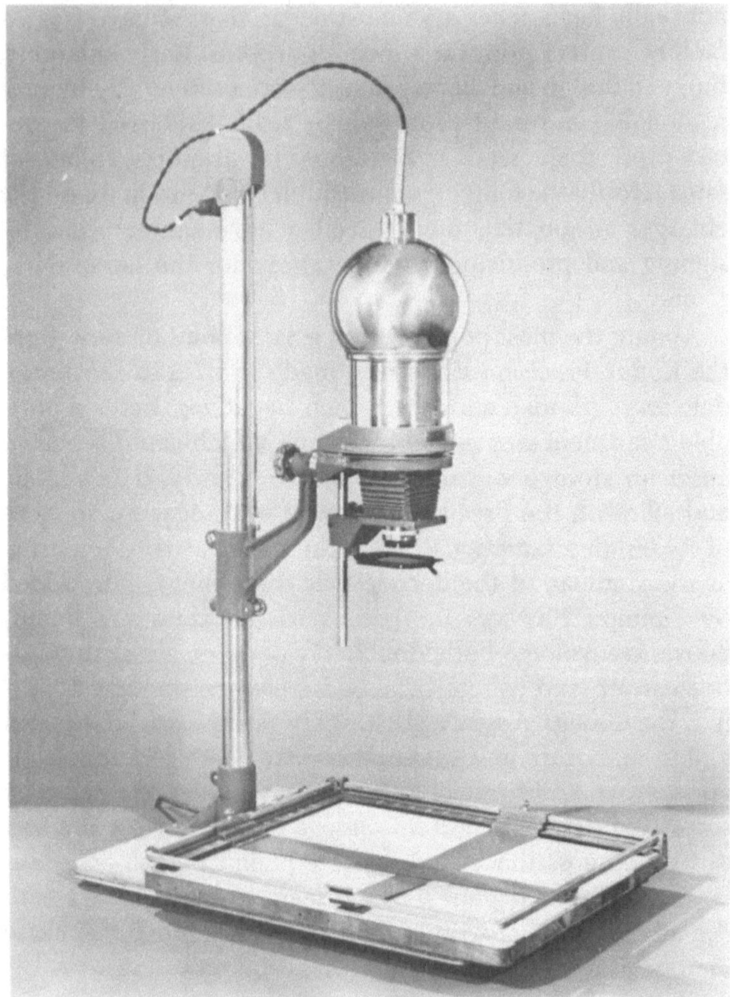

Designed by J. W. Gillon, the Kodak Precision A enlarger represented one of the better units for the serious amateur. Its many accessories gave it wide versatility and respect.

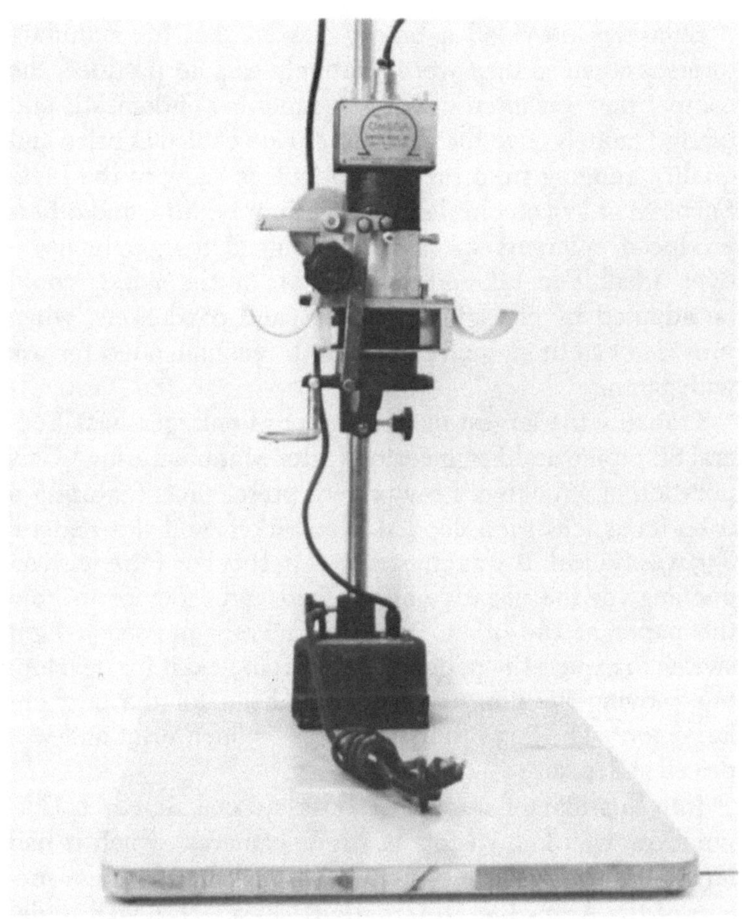

The Cadillac of the prewar enlargers was the Omega A, introduced to 35mm fans in 1938 by Simmon Brothers, of New York. A thoroughly professional darkroom instrument, the Omega A looked and operated very much like the Omega line of today, proving that quality and careful design are not dated by the passage of time.

Enlargers provided a bright new market for manufacturers, and since they were relatively easy to produce, the competition was intense. The large number of domestic and foreign models gave the amateur a wide choice in price and quality, ranging from the Minopticon at $9.95 to the Leitz Focomat at $75.00 plus lens. Kaler, Sunray, Min, and others produced enlargers specifically designed for 35mm negatives. Ideal, Federal, and many others on the market could be adapted by changing the lens (and condensers, when provided) to the 2-inch focal length recommended for use with 35mm.

Probably the largest manufacturer of enlargers was Federal Stamping and Engineering (later Manufacturing) Corporation of Brooklyn. Low-priced, preset units featuring a fixed-focus lens soon flooded the market, and the Federal 835 was typical. It was constructed in shoebox form with an opening for the negative at one end and a device to hold the paper at the other. The user merely pressed a light switch to expose the paper and then removed it for developing, moving the film one negative ahead for the next enlargement. The 835 produced a 3½ × 5-inch print and was priced at $12.50.

Into this market also came International Research Corporation, well known for its Argus cameras, which it had introduced in 1936. Slightly taken aback by the huge success of the Argus line, International Research began to develop what it referred to as a "system" of photography. Part of the system was a means of producing enlarged prints from minicam negatives, an approach copied from the success of the Leica in integrating accessories and equipment with the basic camera. Each owner of a Leica or Argus camera was a potential customer for other parts of the system. Though other American camera manufacturers (notably Universal) also embraced this sales concept, none carried it out with as much inventiveness as International Research. While others were producing the line of accessories one would expect, the Argus line was unique.

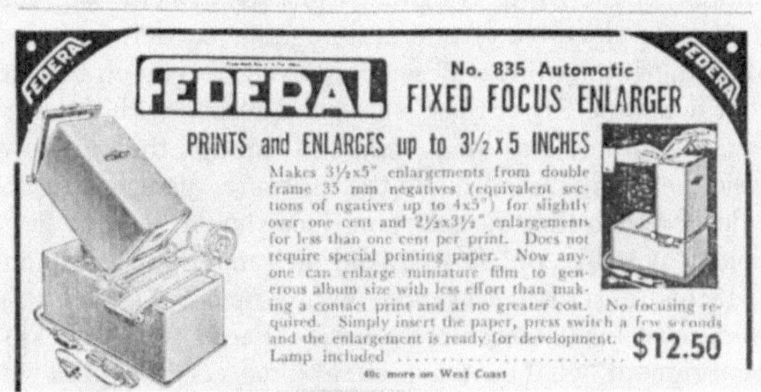

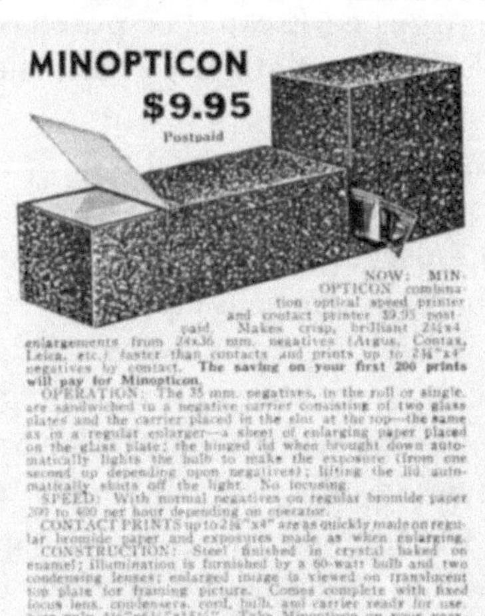

The Federal No. 835 Automatic and Leonard Westphalen's Minopticon were but two of many 35mm fixed-focus enlargers popular in the late thirties.

The first Argus enlarger to appear on the market was Model EA, introduced in 1938 at a cost of $14.75. For an additional $2.75 the purchaser could buy Model EAE, which incorporated an easel in the baseboard of the enlarger. Model EAE was designed to accommodate two condensers and either the Argus A or the Argus AF camera body and lens as its objective system and negative carrier. In its advertising International Research played heavily on the fact that when the EA or the EAE was equipped with the camera, the picture was being enlarged through the same precision lens which had made the exposure, disregarding the rule of optics that an enlarging lens should have as flat a field as possible. But E. Leitz did the same thing, reasoning that the Leica's lens was more highly corrected and sharper than most available enlarger lenses. Faced with that kind of argument, what customer could disagree? For those who wished to spend $21.75, the Argus EAE could be purchased complete with an *f*/5.6 triplet anastigmat lens, which made it unnecessary to use the camera lens.

A few months later the Argus EFA Electromatic Speed Printer enlarger appeared on the market, priced at $35. It was also a preset box-type enlarger, but with an unusual feature: it contained two photoelectric cells and an electric circuit which made use of a radio tube to measure the density of a negative automatically and expose it accordingly. After inserting and positioning the negative and placing a sheet of precut paper in the holder, the user made only two simple adjustments, setting the time for the grade of paper (hard, medium, or soft) and the tone-selector level for the degree of contrast desired. Pushing the "on" button activated the unit. The EFA made 2¾ × 4¼-inch prints on Bromex paper, furnished by International Research in pack-

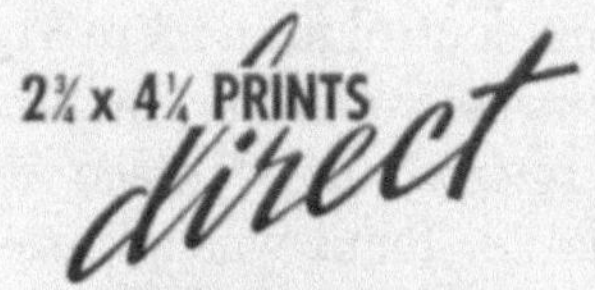

FROM 35 MM FILM
WITH
argus
SPEED PRINTER

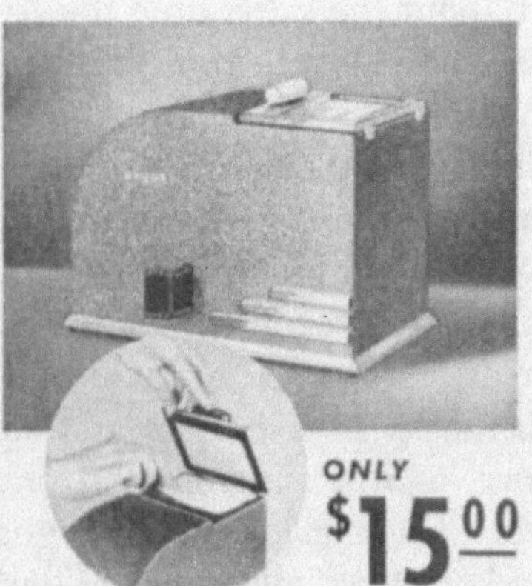

● *No adjusting! No focusing! No trimming! No clutter!* Own this inexpensive ARGUS Speed Printer to make your own enlargements at minimum cost . . . faster than contact printing. *GUARANTEED ARGUS BROMEX PAPER* establishes a reasonable new standard of print size—Bromex is ready-trimmed in perfect proportion to 35 mm negatives . . . glossy or mat finish. Makes generous-size, record prints at no greater cost than contacts. Use Bromex with ANY enlarger for prints of superior photographic quality.

INSERT FILM—PLACE YOUR BROMEX PAPER—A FLICK OF THE FINGER GIVES YOU A 2¾ x 4¼ ENLARGEMENT.

MAKE YOUR PICTURES LIVE ON THE SCREEN . . . LIFE SIZE . . . FULL COLOR

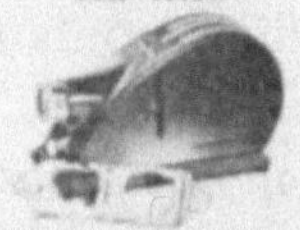

ARGUS 100-WATT PROJECTOR, Model CP . . . $15 . . . High quality, color-corrected 4" focal-length lens. Heat-resisting case. (With carrying case, including indexed slide compartment . . . $19.50.)

Model B . . . POWERFUL tri-purpose film-and-slide PROJECTOR . . . 5" focal-length lens . . . triple condensing system . . . Magazine for 500 single or 250 double frames. (With case and carrier, $35.)

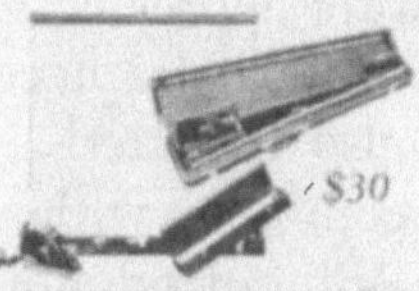

ARGUSCREEN KIT . . . CP PROJECTOR, SLIDE BOX AND CARRYING CASE

A clever combination. Adjustable screen 30 x 32, finest material, will not discolor or crack. Gives brilliance, color gradation and depth. (Screen, Case and Slide Box without projector, $15.)

Call at any leading photographic dealer for demonstration.

ARGUS GLASS SLIDE KIT

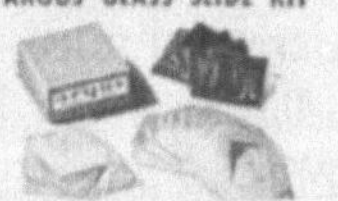

Contains 24—2 x 2 clear slide glass, 12 binding stickers, 24 film-centering masks (sufficient for 1 doz. slides). 75c.

ARGUS BINDER KIT

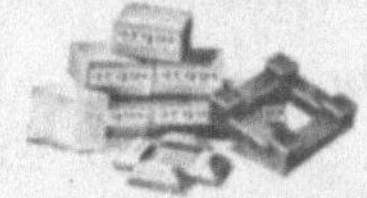

Clever aluminum and sponge-rubber device semi-automatically seals glass slides. With six dozen sets of slides, complete—$5.

INTERNATIONAL RESEARCH CORPORATION, 213 FOURTH ST., ANN ARBOR, MICH.

An Argus product advertisement of 1938.

ages of 36 sheets for 35 cents. The enlarger was also available without the automatic exposure control for a mere $15.

At about the same time the Argus Micro Grain Focusing Easel made its appearance. A 15× microscope eyepiece was mounted on the unit, and the enlarger focus was adjusted until the grain structure of the negative was noticeable. The focusing control was then turned a hair in either direction to prevent the grain structure from appearing in the enlargement. It was recommended that the user focus at the lens aperture to be used to make the print rather than focusing with the lens wide open and stopping it down before exposure, as was the common practice.

For Christmas, 1939, the Argus Jig-A-Mat easel was introduced at $6.75, allowing the darkroom enthusiast to create his own photomontages quickly and easily on 8 × 10-inch paper. The sturdy easel had twelve numbered record levers along one side which were depressed as each separate area was exposed and its cover plate replaced. The Argus Darkroom Light and Clock also appeared, a safelight containing four changeable filters combined with an electric clock. The device sold for $14.50. Without a doubt the most interesting and innovative darkroom ideas bore the Argus nameplate in the days just before World War II.

When one considers the wide variety of unusual darkroom products in the Argus line, it is surprising that Argus received so little profit for them. One reason seems to have been their quality, particularly of finish. The ones we have seen in recent years show ample evidence of slipshod workmanship, unusually poor finish, and lack of attention to detail. The automatic features of the printers never worked well, and when better products filled the marketplace, Argus took the hint and returned to its profitable camera line.

These were only a few of the rather large variety of enlargers and darkroom equipment made available to the 35mm fan, and while only the Omega and Leitz have survived both in concept and in name over the years, there is little doubt that the last few years of the thirties were the most prolific, and in some ways the most inventive, era that the amateur darkroom has ever known—and the excitement revolved mainly around the 35mm format. With the possible exception of the Kodak Flurolite and other cold-light enlargers introduced after the war, creativity on the part of the darkroom equipment manufacturers died rather abruptly.

In the forties and fifties Federal, Solar, Ideal, F-R, and many other enlarger manufacturers brought forth a number of inexpensive models designed for 35mm, but nearly all ended in bankruptcy.[7] In a variation of the old cliché, old manufacturing dies seldom die; they are usually passed on to some other company which puts them back into use. When Federal went under, a St. Louis–based photographic distributor (Arel) acquired the dies, along with those for the Lott rotary drier, and established Arelott to continue making both the enlargers and the driers under the Lott

[7] The Elwood Pattern Works, of Indianapolis, made a neat little 35mm model but specialized mainly in larger sizes—their main advantage. Elwood enlargers were made like tanks and, while heavy and bulky, were practically indestructible. Today apparently the best-selling domestic enlarger is the Testrite, whose manufacturer celebrated its fiftieth anniversary in 1969 and appears to have survived by remaining a small familylike company churning out products others abandoned, along with lower-cost versions of popular items. Its present line of enlargers seems to have evolved from the prewar F-R line, except for one 35mm model. The sole survivor of the Sun Ray brand, the 35mm model is known as the Zenith and is interesting primarily because it was designed for military use and was originally packaged in a portable field darkroom with the Bolsey B Special camera.

brand name. But even the Lott drier had originally been the Warren.

The thermostatically controlled electric drier evolved in the thirties to replace the older hand method of ferrotyping finished prints, which were dried by placing the ferrotype plates in a current of moving air (usually a fan or outdoors in the breeze). While the early driers also used the ferrotype plates, the application of heat made drying a much more rapid process, and the prints dried more evenly because of the canvas aprons which held the prints against the gently curved ferrotype plates. The earliest of the driers were turnover models (still sold and used today), but if one could afford a rotary drier like the Lott, it added an extra degree of professionalism to the darkroom by its very presence. Many sales of rotary driers were made by virtue of their built-in snob appeal. With a developing tank, an enlarger, and a drier, there was very little standing between the amateur darkroom enthusiast and salon-winning prints, other than time and talent—and all of us knew that we had quite a bit of both.

Part II The American 35mm Miniature Camera

8 In Perspective

The remainder of this book is devoted to 35mm cameras produced in the United States from the introduction of the Model A Argus in early 1936; through the last of the American-made 35mm cameras, the Kodak Motormatic 35, introduced in 1960; to the demise of the Argus C-3 in 1968. In the following discussion we have tried to be all-encompassing of the genre by being selective in the companies and cameras represented; readers familiar with this era will realize that so many cameras were designed, manufactured, and marketed in those 32 years that we could not include them all. Thus we have chosen from among the famous, near-famous, and obscure in an effort to give a flavor of the era. In all cases we were able to find models of the cameras discussed, and we have either owned or used most of them. (Our only regret today is that we did not keep all the cameras we have owned at one time or another.)

Unless otherwise stated, opinions concerning quality and handling characteristics are ours and thus necessarily subjective. But in an effort to counter many personal prejudices, we conducted interviews with camera designers and professional photographers to obtain additional viewpoints. All prices quoted are those at which the manufacturer listed his product and are not necessarily the over-the-counter prices.

Since the American 35mm camera was inspired by the success of the Leica and the Contax, it is only reasonable that we begin by briefly discussing those world-famous instruments. In 1935 *The Camera* published introductory reviews of them, and, with the permission of the Ziff-Davis Publishing Company, we reprint below the text of each review. Few modern writers could improve on the descriptive quality of these discussions. The comments also give the reader an idea of the conciseness and clarity with which technical information was presented in those years.

THE LEITZ "LEICA" CAMERA

Compact, accurate, comprehensive, economical—adjectives applicable with added superlatives to a few of the finely-engineered miniature cameras of today, among which the Leica is a featured example. Fitted with automatic distance-measuring, eye-level view finding, focal plane shutter, high speed lens, the whole compressed to pocket-size space, truly the Leica deserves every whit of its well-earned prestige.

All Leica models are rectangular with rounded ends, body covered in black leather-like vulcanite. The Model F is supplied in satin-finish chromium on all exposed metal parts. Weighing 20½ ounces unloaded, it is 5½ in. long, 2⅝ in. high, 1¾ in. thick (140 × 67 × 45 mm.) when closed. To use: lens tube is drawn outward ¾ in., rotated clockwise to lock at infinity focus, nearer focusing by level actuation of helical-threaded lens mount, visually guided by black numerals on fixed tube collar from 100 to 4 feet. Range-of-sharp-focus numerals on adjacent collar automatically indicate depth of focus at any diaphragm stop; collar rotating with focusing tube.

Using standard 35mm film in 5 ft. 10 in. lengths, one spools for insertion in removable metal magazines, 36 exposures are possible at one loading. Picture area, 1 × 1½ inches (34 × 36 mm), 2:3 proportion. To open camera, one base plate key is given ½ turn to remove, whereupon magazine and spool are slid out. Partially-exposed strip removable in dark-room for immediate processing; fully-exposed strip removable in daylight, after re-winding back to delivery magazine. Fresh magazines loadable in daylight. Marketed spools or bulk film may be used.

The right-to-left travel cloth focal plane shutter is wound by 11/16 in. knurled knob, simultaneously transporting film to next exposure, preventing double-exposure errors. Exposure count auto-indicated on beveled collar surrounding winding knob. Normal shutter speeds, 1/20th to 1/500th second, controlled by lift-and-rotate knob on top, slow speeds, 1/20th to 1 second and Time, by rotating knob on front. Shutter-release button on top, threaded for flexible cable attachment if desired.

Built-in optical range finder direct-connected with lens focusing lever, supplying sharp focus automatically without recourse to footage-numerals lens collar. Optical view finder is near lens center, combined with range mechanism; slot provided for 90-degree finder attachment; cast eye-lugs on body for attaching 32-inch leather neck strap; strap and lockable cable release are included equipment, all packed in 15 compartment negative filing case. Standard lens equipment; Leitz Elmar 50mm. *f*/3.5 in flush-surface mount with detachable metal cap; diaphragming from *f*/3.5 to *f*/18. A large variety of additional interchangeable lenses of different foci and speeds and special-duty camera accessories are commercially available, all listed in free literature supplied. The Leitz Leica is manufactured by Ernst Leitz of Wetzlar, Germany, and is imported and distributed in the U.S.A. by E. Leitz, Inc., 60 East 10th Street, New York City.[1]

[1] Reprinted through the courtesy of *Popular Photography*, Ziff-Davis Publishing Company.

THE ZEISS "CONTAX" CAMERA

The world-wide fame of Zeiss in optical and engineering circles bespeaks for the Contax Camera a careful study of its watch-like precision, and at every point, the physical evidence of high standards attained in keeping with the Zeiss reputation. Acknowledged as a leading instrument in the miniature camera field, it is precise in operation, diminutive in size, comprehensive in application, low cost in operation.

The Contax is rectangular in shape with squared ends, body flush-covered in black leather, exposed metal parts in black enamel and bright nickel. Weighing 25 ounces unloaded, it is 5½ in. long, 3 in. high, 2⅛ in. thick (140 × 76 × 50mm.) when closed. To use: lens tube is drawn outward ¾ in., rotated clockwise to lock at infinity focus, nearer focusing from 100 ft. to 3 ft. by finger rotation of knurled disc in camera top, actuating helical-threaded lens tube with white footage numerals on surrounding collar. Black numerals on adjacent fixed collar give depth of focus range at any diaphragm aperture used.

Using standard 35mm film in 5 ft. 10 in. lengths, spooled for insertion in removable metal magazines, 36 exposures are possible at one loading. Picture area 1 × 1½ inches (24 × 36 mm.), 2:3 proportion. To open camera, two base keys are given ½ turn, whereupon the one-piece back, base and ends-section slips off to expose entire film channel and spool chambers. Marketed spools or bulk film may be used, partially-exposed strip may be removed for immediate processing, two magazines may be inserted for daylight removal without re-spooling. Contax spools permit daylight removal without use of second magazine.

The metal-leaved curtain downward-travel focal plane shutter is wound by a ⅞ in. knurled knob on camera front. Shutter settings, from bulb and ½ second to 1/1000 second, obtained by rotation of lower dial to segmental aperture, followed by lifting and rotating winding knob. To eliminate double-exposure errors, shutter winding knob simultaneously transports film for next exposure. Release button on top is threaded for reception of cable release. Exposure counting automatically indicated on ⅞-inch beveled dial adjacent to release button, dial manually resettable at will.

The flush-surface built-in optical range finder uses entire body length for base, range and view-finder eye-pieces at extreme left end. Slidable aperture plates mask view finder to match varying lens foci. Clip is fitted to camera top for reception of special finders, etc., two swinging eyes for 30-inch neck strap are attached; strap and cable release supplied. Standard lens equipment; Carl Zeiss Tessar 50mm. *f*/3.5. in flush-surface mount with metal cap; rim-set diaphragming from *f*/3.5 to *f*/22. Zeiss Proxar supplementary lenses, wide choice of interchangeable lenses in Contax mounts and camera accessories are available. The Zeiss Contax is manufactured by Carl Zeiss of Jena, Germany; imported and distributed in the U.S.A. by Carl Zeiss, Inc., 485 Fifth Avenue, New York City.[2]

[2] Reprinted through the courtesy of *Popular Photography*, Ziff-Davis Publishing Company.

9 It All Began with . . .

The Leica preceded the first commercially successful American 35mm camera by a decade. The Argus dates to 1931 and originated in the depression-ridden heartland of the United States. A group of Ann Arbor, Michigan, businessmen led by Charles A. Verschoor pooled their talents in an effort to boost their community out of the economic doldrums. Their idea was to create new jobs in a growth industry which would step up employment in Ann Arbor. Radio seemed to be the perfect answer; it was enjoying a firm hold on the American public. Verschoor formed the International Radio Corporation to manufacture inexpensive radios, using a molded-plastic case to keep costs low. The International Radio Corporation was successful from the beginning. Its initial product was the Kadette, the first five-tube ac-dc mantel radio on the market, and its patents were held exclusively by IRC. Business was so good that Verschoor soon added the International model and even ventured into the private-label radio business with gratifying financial results.

By 1933, IRC was booming, and its contribution to Ann Arbor had become substantial. It had created new jobs, which had also brought additional money into the city, and, even more important, it had generated team spirit in the business community—Ann Arbor was on its way, and IRC was in the lead. But radio manufacture had one major drawback. It was a seasonal industry, with an annual slack period in late spring and early summer. Verschoor reasoned that if a product line could be developed which relied heavily on the use of plastics the facilities of IRC could be better utilized and its production line and sales force could be kept busy the whole year. On a trip to Europe, Verschoor had discovered the Leica camera and had been impressed by its increasing popularity. Universal Camera Corporation was just beginning to hit pay dirt with its line of inexpensive UniveX cameras, and Verschoor believed that International Radio could do the same with a 35mm camera, provided the price was low enough and the appeal wide enough to attract a large market. The use of plastic could help keep the cost reasonable. At Verschoor's behest Gustave Fassin worked out a simple design in 1934. Priced at $9.95, the camera appeared on the market early in 1936 as the Model A Argus.

Publicity had been set into motion long before delivery to dealers, and the Argus A was an instant success, selling well over 30,000 units in its first few weeks on the market. The price soon went up to $12.50. Like Universal before him, Verschoor had confounded the experts who said that it could not be done, though even after he had proved them wrong, they maintained that the camera's popularity would not last because of the small negative size. But Verschoor felt vindicated by the sales pattern of the Argus, and, believing that the potential of the line was far beyond that of radios, he committed IRC to the future by selling its radio patents to the Radio Corporation of America. The International

Radio Corporation was renamed the International Research Corporation, and plans were laid for extending the success of the Argus A into other photographic areas.

The outgrowth of several economy measures, the lightweight (14-ounce) body of the Argus A was molded from Bakelite. The film-wind and rewind knob, exposure counter, and lens mount were made of metal and were firmly attached to the body. The detachable camera back was a formed metal plate of stamped steel (later of aluminum) with a satin finish, which added a touch of class to an otherwise stark appearance. The back snap-locked in and out of place, giving quick access to the interior. Economy touches

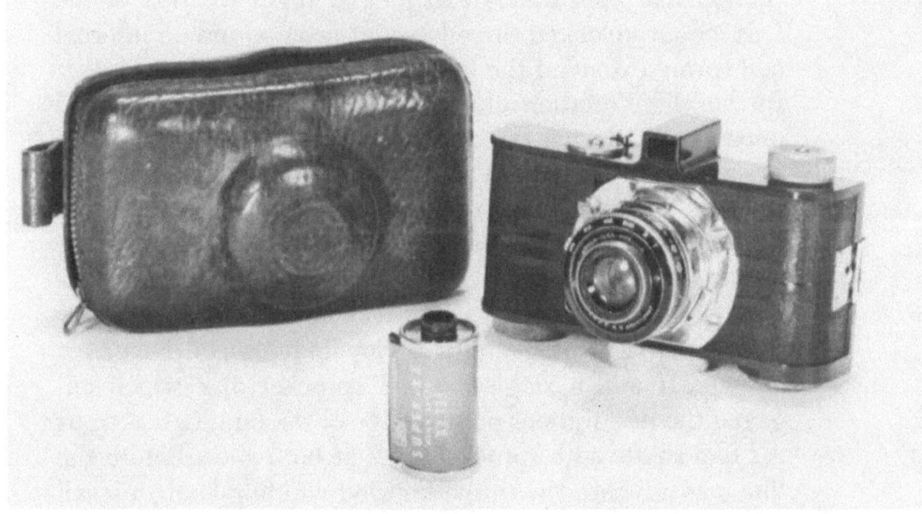

Shown here with its semisoft-pouch case, the original Argus Model A made 35mm photography a popular hobby with millions. One of the first produced, this example looks almost exactly the way it appeared on a dealer's shelf in 1936.

inside the camera were apparent in the sponge-rubber bases; the film pressure plate was firmly mounted on one base (later made free-floating), and another held the film cartridge in place after loading. Interestingly enough, had it not been for Kodak's introduction in 1934 of the daylight-loading cartridge for use in its Retina, the Argus A could never have been designed and manufactured to sell for such a low price.

While the Argus A was virtually a fixed-focus camera, it made use of a collapsible lens tube which facilitated carrying and allowed for a two-position variation on the traditional fixed-focus lens, giving the camera a somewhat more professional appearance and feel. A slight rotation of the lens mount unlocked three bayonet prongs, and an internal coil spring extended the lens to a position of infinity focus; further slight rotation allowed the lens tube to move forward once again, altering the camera's focus to subjects within a range of 6 to 12 feet.[1] The Argus A would not take sharp pictures unless its user extended the lens tube correctly for the distance from the camera to the subject. One of the authors invariably neglected to do so, and the result was roll after roll of fuzzy negatives. The large exposure-counter dial at the right of the optical viewfinder was driven by a brass gear and a single-toothed sprocket disk which engaged the perforations on one edge of the film. Later Argus A's had shafts with sprocket disks at both ends. Before the film was advanced with the knurled winding knob, a small release plunger was momentarily depressed to free the automatic-stop mechanism on the film transport. Failure to depress it meant ripping the film's sprocket holes.

The heart of any camera is its lens and shutter. The Argus

[1] The range was soon altered to 6 to 18 feet.

A incorporated an Ilex combination designed especially for its use: the 50mm IRCF *f*/4.5 anastigmat lens mounted in a self-cocking Argus-Ilex-Precise shutter. Although it had four diaphragm settings from *f*/4.5 to *f*/11 and four shutter speeds designated at 1/25 to 1/200, *T* and *B*,[2] the suspicion that its rated aperture was not *f*/4.5 was widely held in the trade and has lingered through the years. Interested persons still maintain that bench testing of the Argus A's lens proves it to be about *f*/6.3 in aperture, with a shutter correspondingly slower than marked to compensate for the approximately one-*f*-stop difference between the real and the marked values.[3] It appears that this lens and shutter cost IRC $1.

The Argus A proved to be a functional camera, easily held and used. It was light enough to carry handily, and the collapsible lens tube (and Argus advertising copy) encouraged the owner to slip the camera into his coat pocket (something that proved a bit difficult to do with the average coat). Fairly well balanced and shaped to allow a firm but comfortable grasp, the Argus A (available in a black, tan, or gray body) had enough appeal and appearance to convince prospective buyers that it was an instrument for serious picture taking.

The Argus A was manufactured from 1936 to 1941. In 1937 the AF and B models appeared.[4] They were identical to the A in appearance, but the AF had a complete focusing range from 15 inches to infinity instead of the two-position

[2] While some Argus A's had diaphragm settings of f/4.5, 5.6, 8, and 11, others used the German system *f*/4.5, 6.3, 9, 12.7, and 18. The top shutter speed also varied; on some cameras it was 1/200, on others 1/150.

[3] The company later made pointed advertising claims that its camera lenses were exactly the rated speed.

[4] The AF was manufactured in 1937 and 1948, the B only in 1937. The B was the only Argus of this era that had a foreign-made shutter.

tube focusing of the A. The B, which sold for $25.00, had a 50mm *f*/2.9 anastigmat triplet lens in a Prontor II shutter, giving a range of shutter speeds from 1 to 1/175 second, *B* and *T*, and incorporating a built-in self-timer. The A2 ($12.50) and A2F ($15.00) appeared in 1939.[5] They were simply the A and AF with the addition of an extinction meter with a sliding calculator on the top, at the left of the viewfinder. The prewar A series was completed with the introduction in 1940 of the A3, the basic model from which the Argus Colorcamera[6] came, and the AA, a flash-synchronized (for the C-3 flashgun) version of the original Argus A, with a 50mm *f*/6.3 fixed-focus anastigmat triplet lens and two shutter settings, instantaneous and time (*I* and *T*). Thus the consumer was faced with a rather bewildering array of Argus A cameras during the late thirties and early forties, and even today it is hard to tell them apart. The picture soon became even more cloudy, as we shall see.

While the Argus A put the International Research Corporation in business, photographically speaking, it was to be the next series of cameras which would make Argus a household name and assure its prosperity for well over two decades. The Argus C was introduced in 1938 and went on to achieve a success which no other American 35mm ever enjoyed. Over 2 million of the "bricks," as the rectangular body was referred to in the trade, were produced during the next 28 years, and the popular C designation was extended after World War II to incorporate another body design evolved from the postwar Argus Markfinder 21 via the Colorcamera of prewar vintage.

[5] The A2 was manufactured from 1938 to 1950, the A2F from 1939 to 1941.

[6] The A3 was manufactured from 1940 to 1942, the Colorcamera in 1940 and 1941.

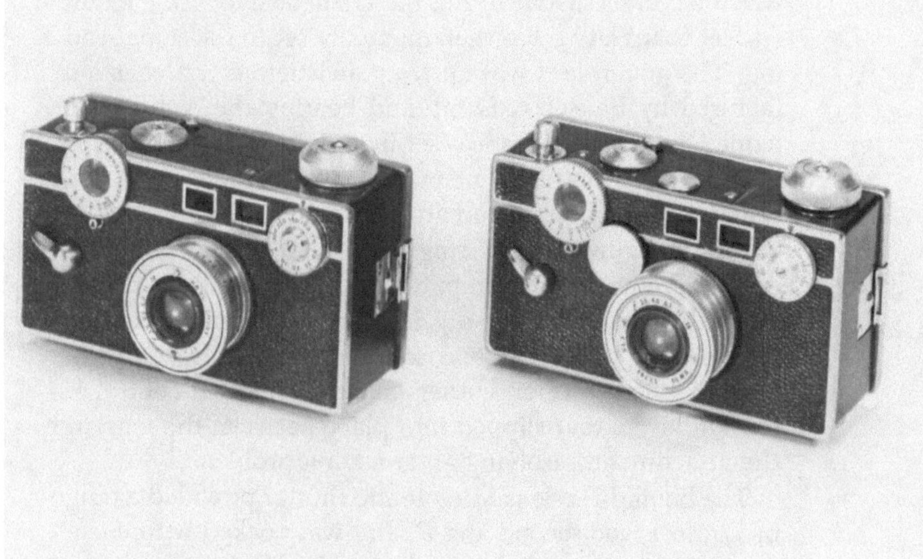

The Argus C (left) had a noncoupled rangefinder; in the C-2 (right) the rangefinder was coupled to the Cintar lens. The earliest C also had a slow-speed selector below the cocking lever. Both cameras had a 1/5-second shutter speed, which was later dropped from the C line.

Also designed by Gustave Fassin, the Argus C went into production early in 1938, reaching dealers' shelves in March with a $25 price tag. The metal-and-plastic body incorporated a built-in optical rangefinder of the split-image sextant type. Looking through the rangefinder window, the user selected a vertical line in the subject, turning a calibrated knurled wheel until the two segments of the line appeared as one. The reading on the wheel nearest the indicator arrow was the correct distance, and the C's helical focusing mount (3 feet to infinity) was then manually set to the same reading. The 50mm lens was an *f*/3.5 anastigmat triplet manufactured by Bausch & Lomb and bearing the Argus trade name Cintar.[7] The lens was set in an interchangeable screw mount. The front-element mount was threaded to accept special screw-in filters and attachments, instead of the slip-on or set-screw adapter rings prevalent at the time. The diaphragm was also set from the front, a frustrating arrangement that made it virtually impossible to change settings once a filter or a lens hood was secured in place. In time Chess-United, Inc., and others made diaphragm-control extension levers that slipped into place between the lens and the attachment, resolving an annoying problem.

The behind-the-lens Micromatic shutter provided a range of 1/5 to 1/300 second and *B*, and was cocked before each exposure by a lever below and at the right of the calibrated rangefinder wheel. Initially, the Argus C also had a fast-slow shutter selector switch placed below the cocking lever, but this function was soon incorporated internally, and today an Argus C bearing this switch is extremely rare. The early models were highly trouble-prone in almost every respect, but their primary problem was a vignetting of the lens—the lens did not cover the negative, and there was a resulting darkening effect at the corners of the negative. Before the year was over, the C-2 made its appearance with the rangefinder coupled to the lens, an improvement that eliminated

[7] Bausch & Lomb produced the Cintar lenses for Argus under contract for a while, but Ilex made most of them. Other Cintars (of poor quality) were made by Graf Optical Company, and Argus itself made many after absorbing Graf in 1939. While owners seldom had anything but praise for the Cintar, photographic magazines were often harsh in their evaluation of what they termed a "fair" lens.

some of the bugs. Turning the rangefinder wheel to align the two images also turned the lens mount, automatically focusing the lens for the correct distance, now a range of 3½ feet to infinity. The C-3 followed in October, 1939. It was identical in function and appearance to the C-2 except for built-in flash synchronization through a plug-in flashgun at the left of the camera. It was this model (initially selling for $35) which accounted for most of the 2 million sales of Argus C's.

The camera had polished surfaces on the top, bottom, and ends. The front and back of the molded black-plastic body were covered with black leatherette. A cast-aluminum frame (soon to be replaced with a die-cast body and chrome plating) of the front panel did its best to disguise the ugly squatness of the heavy (24 ounces) beast, and yet the public seemed to be captivated by its scientific appearance.[8] The unsightly brute was off to set an all-time sales record in its class.

By and large the C-3 proved to be an extremely reliable performer. Its main problem (other than a loosening of the tripod socket) was its rangefinder. Virtually every C-3 one of the authors (Lahue) sold in his years behind the camera counter was eventually returned to the factory for internal adjustment.[9] People also continued to rip out film perfora-

[8] This surprising fact was turned up by informal surveys taken by competitors who queried dealers, salesmen, and customers in an effort to determine the secret of its success.

[9] The rangefinder assembly was mounted on a mechanical plate, which was then inverted for insertion into the camera. The mechanical plate was a thin piece of metal which formed part of the top of the camera and was so positioned that, when the user grasped the C-3 firmly enough to deliver the 7/10-ounce release pressure necessary to trip the shutter, the pressure flexed the mechanical plate. Repeated use tended to throw the rangefinder mechanism out of line. Argus received approximately 50,000 complaints a year about the C-3 rangefinder.

tions because they did not push the button release before winding, but that could hardly be considered the fault of the camera.

The word-of-mouth advertising of this rugged camera was probably more important to its sales than any advertising campaign could have been. When new owners were asked why they had chosen the C-3 over competitive cameras, their reply was invariably the "sharp, beautiful color pictures" which their friends had taken and the impression that the camera was "built to take care of itself." Housed in their well-padded leather carrying cases, thousands of C-3's took the rough treatment of their owners in stride, performing like workhorses over the years. The C, C-2, and C-3 were not particular delights to hold and use, but the design proved to be neutral in that, while one seldom fell sufficiently in love with the camera to fondle and pamper it, the brick was not too awkward or clumsy to use efficiently.

The last of the prewar Argus cameras were the K and M models, introduced in 1939; the A3, introduced in 1940; and the CC, or Colorcamera, introduced in 1941.[10] A most unusual camera, the Argus K was a commemorative design in honor of the one hundredth anniversary of Daguerre's presentation of his process to the French Academy of Sciences in 1839. The all-metal body of the K featured a top-loading

[10] The K and M were discontinued in 1940, the A3 and CC in 1942. During the war, however, the CC was made for the army without the meter and finished in black. Featuring a front-focus *f*/4.5 lens, it is an unlisted Argus model which does not appear officially in the company history. IRC announced a Model D automatic-sequence camera at $24.50 in 1939 which was to use an *f*/4.5 lens and a shutter with speeds up to 1/200. The camera was designed to compete with the Robot, made by Otto Berning (later the Berning Company, of Düsseldorf, Germany). Tooling was completed, but the camera could never be debugged, and the dies were used in manufacture of the Model K. This model proved troublesome and unpopular and was also a financial failure.

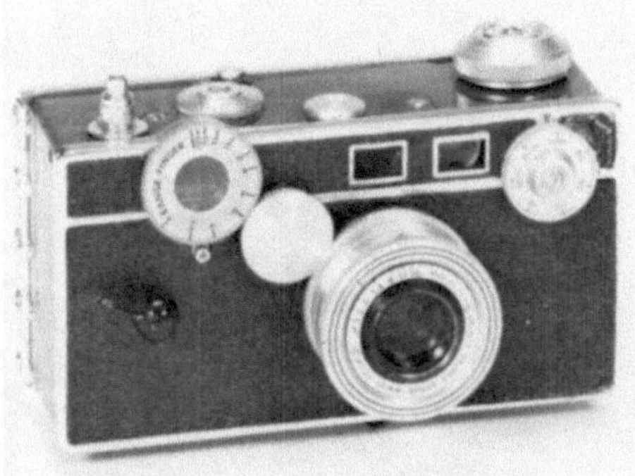

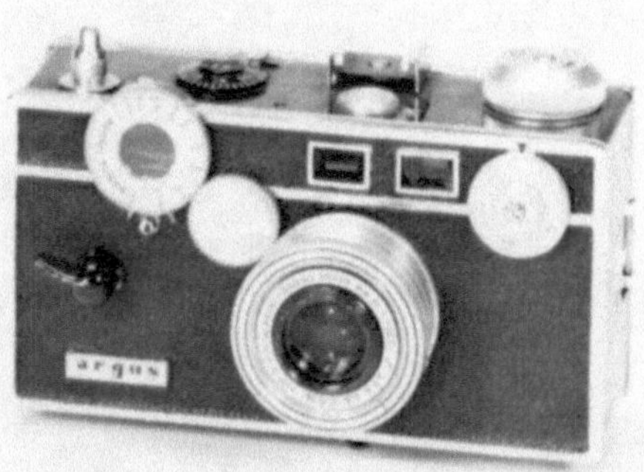

While usually considered one camera model, the Argus C-3 was regularly improved on throughout its production life. Here are four examples which show what Argus did, at least on the outside, to make the "brick" more desirable. Minor changes—the addition of an accessory shoe and an Argus nameplate and the elimination of some speed markings (right)—were the first major alterations.

system, an *f*/4.5 color-corrected lens, and a shutter rated from 1/25 to 1/200, *B* and *T*. Focusing was 3½ feet to infinity, with its scale color-keyed in three segments for easy zone focusing. This rather unique experiment also made use of a built-in coupled extinction exposure meter. The user preset a series of dials on the camera back, looked through a small peephole at the bottom left center of the camera, and moved the aperture indicator until no light could be seen. He then shifted his eye farther left to the viewfinder window to frame the picture and tripped the

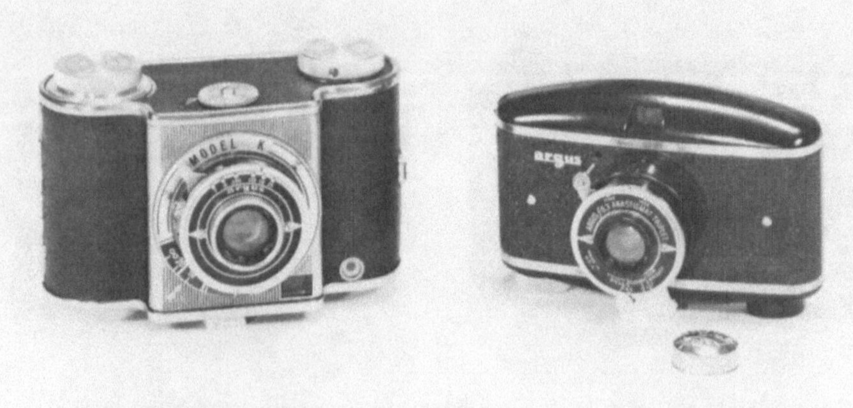

The Model K (left) is one of the rarest models of the Argus line; the camera was on the market for less than a year. Manufactured in celebration of the one hundredth anniversary of Daguerre's process, the camera was a complete departure from other Argus models—and a failure in the marketplace. The Model M Argus (right) was unique in that it took both single- and double-frame pictures on 828 film. It was designed for World War II servicemen. After the war it appeared as the Cameo, which was offered primarily as a premium item.

shutter release. It was an inexpensive ($19.50) innovation which was difficult to understand, clumsy to handle, inefficient in operation, and impossible to sell; it disappeared within a year.

The Argus M was a streamlined box camera with an *f*/6.3 color-corrected triplet lens in a collapsible mount. It took either 12 or 24 pictures on 828 film. Gift-packaged for Christmas, 1939, with case, two rolls of Arguspan film, and

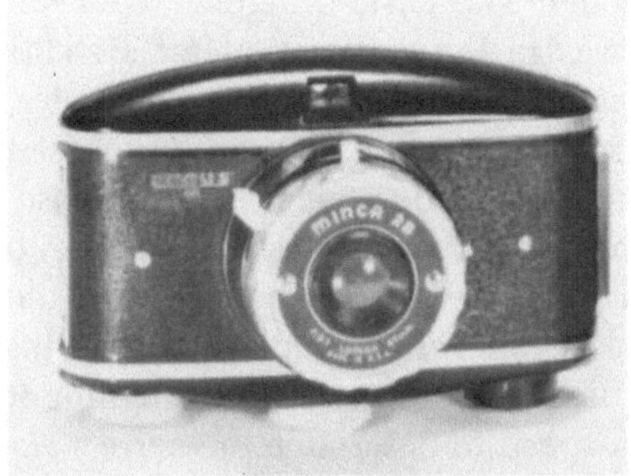

The Argus Model M, which with minor changes was also sold as the Minca 28 and the Cameo, began life as a soldier's camera in 1941. The size was right, and it was economical of film—it would take half- or full-frame Bantam-size pictures in the M version. But since 828 film was all but impossible to obtain during the war, the camera was not very popular. The dies were later sold and a premium camera produced from them. It has become one of the rarer Argus models.

a portrait lens, it sold for $9.95. Although 700,000 to 800,000 Model M's were manufactured, the camera never reached what could be termed great popularity with the public. It was, however, one of the few cameras Argus produced during the war. It was called "the soldier's friend" by the company, a rather strange appellation for a camera using scarce 828 film. Later the dies were sold to a company in Pennsylvania which produced it under the name Cameo and housed the lens in a fixed mount with a three-speed shutter (1/25 to 1/100, *T* and *B*). The camera found its best market as a premium item.

Designed primarily by Ned Graef, with some assistance from Clint Harris, the Argus A3 was the first significant departure from the basic lines (A and C) to enjoy any degree of interest. It contained a host of unusual features, as did the

Colorcamera, or CC model. The basic difference between the two new cameras was in the exposure-meter system: while the A3 retained that old favorite the extinction meter, the CC incorporated a selenium-cell light meter purchased from an outside manufacturer. Made under Weston license, the meter contained a needle and a scale calibrated from 0 to 22. The user pointed the camera at his subject, read the light value at which the needle rested, and transferred his reading to a calculator dial on the camera back, aligning it opposite the proper film speed to determine the correct exposure.

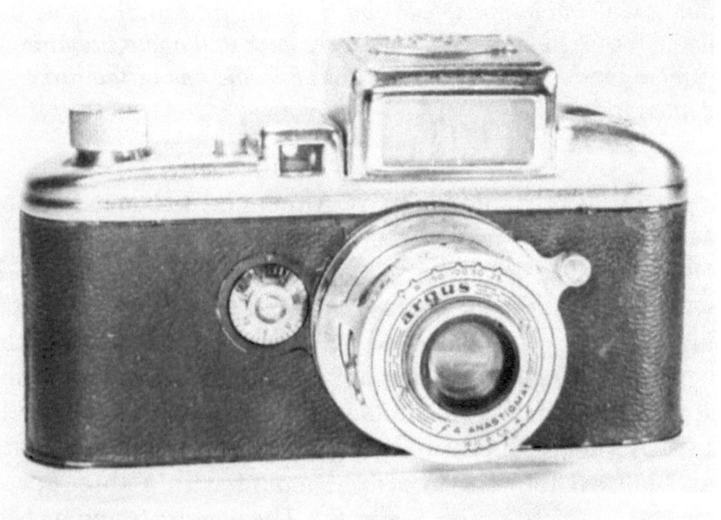

One of the novel American 35mm cameras, the Argus CC, or Colorcamera, featured a built-on selenium light meter. After reading the meter scale, the user transferred the number to a calculator dial on the camera back and then set the lens and shutter accordingly.

Manufactured completely of metal, both the A3 and the CC had *f*/4.0 lenses and self-cocking shutters rated from 1/25 to 1/150, *T* and *B*. The settings were visible from either the front or the top of the collapsible lens mount. On both cameras focusing was accomplished by rotating a lever which moved the lens mount between positions of 15 inches and infinity. The exposure-frame counter was placed on the face of the camera body. Nestled into the top housing was one of the most unmanageable film-rewind knobs ever to confound man. Split into two parts, the top of the knob folded inward to form a V, and rewinding a 36-exposure roll of film not only was a time-consuming, patience-shattering experience but also usually cost the owner a fingernail. The meter's position on the CC interfered with easy use of the viewfinder—a problem not encountered on the A3—but it was an interesting variation from other cameras in the Argus line, and its built-on meter pointed toward the future.[11]

Late in 1938, IRC purchased the Chicago-based Graf Optical Company (a well-known lens manufacturer which had become a division of the General Scientific Corporation of Chicago in September, 1932) and began producing its own lenses. Many camera accessories[12] were developed: a macro attachment for copy work ($8.75), the Techniscope, which served as a copying stand and a microscope adapter,

[11] In 1935, Zeiss had introduced the Contaflex, a 35mm twin-lens reflex with a built-in photoelectric meter. In 1938, Kodak's Super Six-20 (a roll-film camera) integrated the meter and lens to achieve automatic-exposure control. But the Argus CC was the only American-made 35mm to appear with a photoelectric meter as an integral part of the camera during this period.

[12] IRC also marketed its own line of slide projectors, the Model B (manufactured by SVE) and the DP series (manufactured by Spencer Lens Company, later American Optical Company), as well as screens, slide-binding kits, and the complete darkroom line mentioned earlier.

among other uses ($29.50), and the Photar light meter licensed under Weston patents ($8.75). Copy and portrait attachments, filters, and telephoto lenses were among the other accessories made available for the Argus C camera. Independent manufacturers developed and marketed the accessories Argus overlooked, among them the Laub coupled rangefinder for the AF (which fastened to the bottom of the camera) and the Shull Speed-O-Copy attachment for the C-3 (also made in models to fit the Leica and the Contax). IRC sold film and sensitized paper under the Argus brand name and publicized its products widely and effectively. The term *candid camera* became almost synonymous with Argus, and photo magazines which regularly featured articles on miniature camera techniques made extensive use of pictures taken with Argus cameras. Unlike many of its competitors, IRC was serious about both its products and its future and was well managed by Verschoor and his associates. It was a success story in the best American tradition.

Late in 1940, IRC began moving into production for the government, beginning with the lend-lease program. By August, 1942, the company had converted to full production for the war effort. By the end of World War II it had won the Army-Navy E Award five times. Like the Mercury II (see Chapter 10) and the Clarus (see Chapter 16), before the end of the war the C-3 got a head start on its competition by receiving a multimillion-dollar order from the army post-exchange system and priority on materials to resume manufacture. Reconversion to a peacetime consumer market came at war's end, and not long afterward Argus, Inc. (the new name adopted in 1944), made a major mistake in market timing.

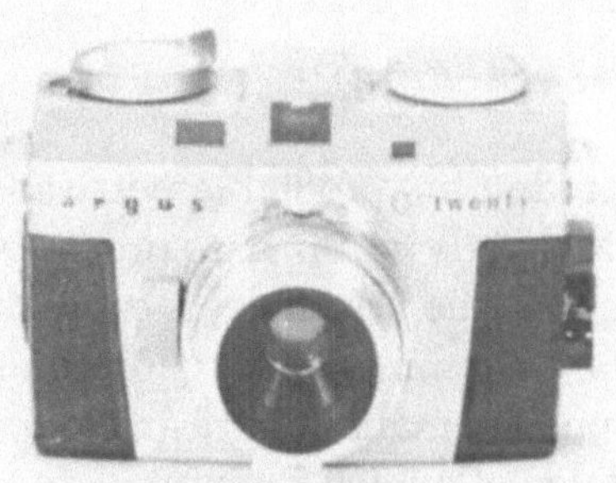

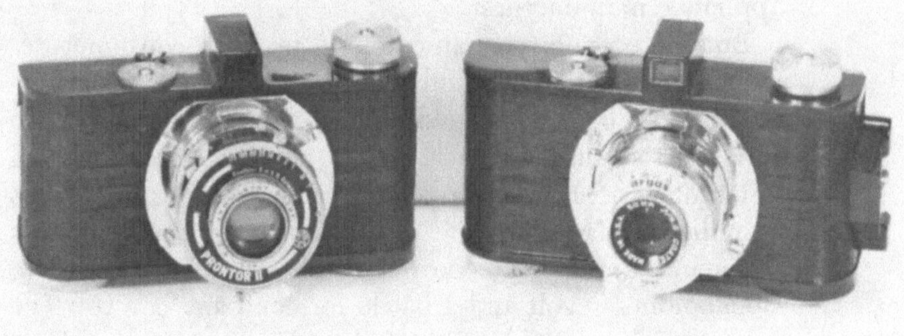

The Argus A with a German lens and shutter (left, below) was the first f/2.9 marketed by the company and was not unlike the Leica B. It was apparently too expensive for the market. The C-Twenty (center, top) was a compact rangefinder camera produced after the war, also with a foreign-made shutter. Although attractive, it reportedly had a poor rangefinder, and it failed to meet competition. With eternal hope built on past success, the manufacturer resurrected the original A body after the war, added a synchronized shutter, and sold it as the Argus FA (right, below) at a "bargain" price of $29.00. It also met with limited success; other cameras provided more for the money. If it had sold for $9.95, the story might have been different.

Argus shared in the profitable postwar boom, as a ration-weary public eagerly grabbed up products almost as quickly as they came off the assembly lines, and watched its sales double from $4.5 million in 1946 to $9 million in 1948. Argus continued to manufacture artillery sighting scopes for the army, sales of which amounted to 25 per cent of its gross business in 1946. In the same year Argus decided to diversify its operations wth the acquisition of a Minneapolis appliance manufacturer.

But the economy began catching up with consumer demand, and, with no brand name on which to rely when competition became rugged, there seemed to be no way to turn the red ink into black. Argus' new appliance division lost money at an alarming rate, eventually costing Argus $1 million to liquidate and turning the 1949 profit-and-loss statement into a $610,000 net loss.[13] This fiasco led to a stockholders' revolt and a battle for company control. The insurgents won and brought Robert E. Lewis from a New York engineering-consultant firm to direct the future fortunes of Argus. Surrounding himself with a host of young men in key positions, Lewis changed the corporate name to Argus Cameras, Inc., and initiated over a dozen new models, phasing out all the former models in the process, except for the seemingly indestructible C-3.[14]

In March, 1950, in what seemed a shrewd move, the

[13] This diversification came at the worst possible time, just before the economic setback of 1948–49. Argus turned the financial corner in 1950 with a $778,000 profit on gross sales of $5.3 million; by 1953 it was grossing almost $20 million a year, with net earnings of over $3 million.

[14] The sales curves of the C-3 were like a Coney Island roller coaster. Sales would dip to an almost-zero level, and Argus would announce the camera's demise. Panicky dealers would then stock up with the remaining supply to ensure stock of a slow but certain seller. To fill these orders Argus would have to schedule another production run, and so it went for years.

company announced at the convention of the Master Photo Dealers Association that drastic price reductions were being made to clear its warehouses. The merchandise was sold below cost, and the sale was highly successful. Imagine Lewis' chagrin a few months later when the beginning of the Korean conflict reversed economic conditions and demand shot up. If the company had only held off a few months longer, the stock could have been easily moved at regular prices.

In the years just after the end of World War II, Argus reintroduced the A2, the C-3, and the Argoflex EM. They were joined by the Model 19 and the company's only new contribution to 35mm design, the sleek Argus Markfinder 21.[15] Introduced in 1947 at $58.08, the 21, of black leatherette and aluminum, was a smartly styled, well-built camera which should have become more popular than it did. Actually, its design was not entirely original; the 21 shared the same body with the Colorcamera of prewar years with but minor exceptions. The bottom and back castings were exactly the same, including the area which held the exposure calculator on the CC. The body casting also appears to have been the same. By clever engineering the shutter of the 21 was built into formerly unused areas of the CC. A rigid mount replaced the former collapsible mount, providing additional space. The camera featured an *f*/3.5 Cintar lens in an interchangeable mount and a behind-the-lens leaf shutter with speeds of 1/10 to 1/200 and *B*. The lightweight

[15] The E was a twin-lens reflex first marketed in 1940 and superseded by the EM in 1948. The former had a plastic body, the latter a metal one. The EM was replaced in 1948, the same year it was introduced, by the EF, an EM to which synchronized flash had been added. The Argoflex was private-branded for Montgomery Ward as the Wardflex. The 19 was an updated version of the prewar Model M, sometimes referred to as the Minca 18 and manufactured only during 1947 and 1948.

chrome lens mount was manually focused to 3 feet.[16] The shutter-speed selector dial was placed at the right of the lens mount on the camera face, and all controls (including the film-wind and rewind knobs) were large knurled rings which made the 21 a delight to use.

Mounted in an anodized-aluminum housing across the top of the 21 was a large viewfinder which incorporated the camera's outstanding feature, the Markfinder. This was the company's first bright-line frame viewfinder, a luminous rectangle with an × at the center and optically imposed by Plexiglas prisms upon the viewfinder image to show the exact area covered by the lens and provide help in framing the picture. The brighter the light in which the camera was used, the more brilliant this superimposed image became, but it was also easily distinguished in near darkness.

The 21 was designed in anticipation of a large military contract (which never materialized), and this adaptation of the cross-hair reticule principle was provided to aid the user under all kinds of lighting conditions. It was one of the most enlightened viewfinder designs of the period.

Departing from the standard Argus plug-in flash unit, the 21 incorporated a hot shoe which was neatly hidden above the viewfinder. A selector switch underneath gave the user a choice of class M or F synchronization, depending upon the flashbulb he used.

Neat in looks (black leatherette on aluminum), carefully finished with such touches as an engraved line on the housing top to indicate the film's focal plane, smooth as silk

[16] Like the manufacturers of the Leica, Argus tried to convince amateurs that they should use the same lens that had taken the negative to enlarge it. The 21's removable lens was manufactured that way for just this reason. No other lenses were ever made for it—nor was an apparently scheduled enlarger.

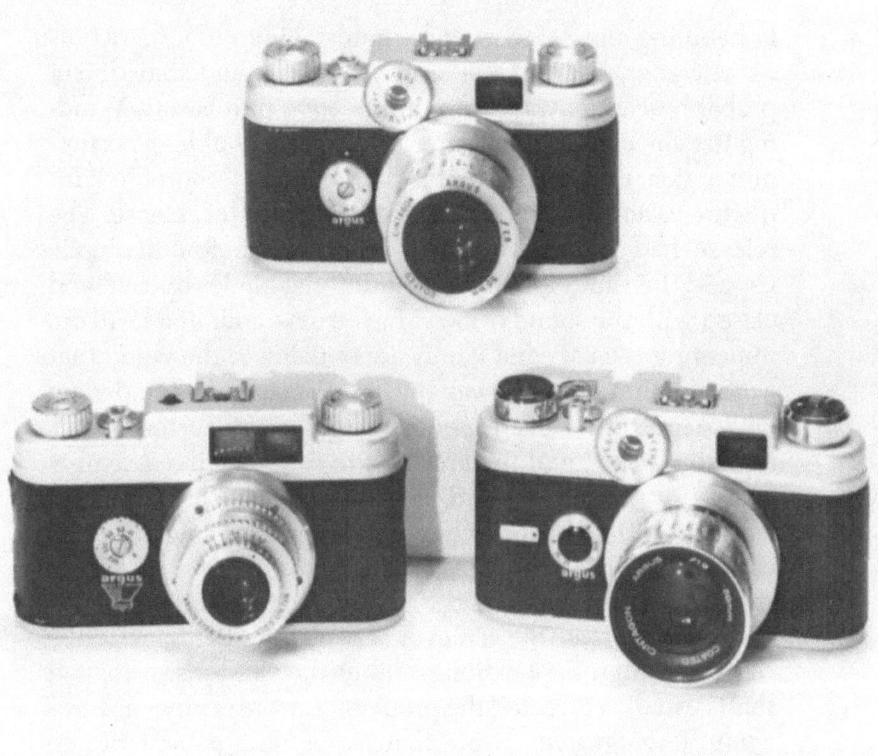

The great postwar hopes of Argus were based on the C-4 series. Beginning as the Markfinder 21 (left) and ending with the C-44R with rapid winder (right), it was the best-looking Argus ever made in this country. The actual bodies were those of the prewar Colorcamera, and so the line was fairly inexpensive to produce. Shutter speeds differed on each camera, although the shutter appears to be identical in each model. Beginning with the C-44 (center), the cameras had interchangeable lenses.

in handling and balance, and comfortable to use, the Argus 21 at $57.25 had but one apparent fault, and that defect probably scared away many prospective purchasers. Winding the film cocked the shutter (a common double-exposure-prevention device), and the photographer composed his picture carefully, gently depressing the shutter release. The release had a lengthy downward travel before it finally tripped the shutter, and when it did, the blades opened and closed with the sound of two Mack trucks colliding head on, unnerving the user and loudly announcing to the world that a candid photographer was at work. While some dampening device must have been feasible, Argus never corrected this disturbing flaw, and the final 21's to roll from the assembly line in 1952 still clacked noisily as their shutters were tripped. The 21's shutter had one other annoying eccentricity: although a sturdy, simple design, it required liberal amounts of lubrication and was prone to freezing up. It was not a fatal defect; the removal of a plate in the back of the camera and an application of oil to the shutter's gears and shafts usually corrected the problem. Burt Murphy, in *Argus 35mm Photography*, wrote that Argus design came of age with the 21. We would add that its great days ended with the passing of the Model 21 in 1952.

The cameras introduced during Lewis' reign appear to have been designed in committee. The C-3 became the Standard C-3 and then the Match-Matic C-3 ($89.50), on which exposure was set by means of EVS numbers taken from a clip-on light meter.[17] This model in turn became the C-33, an adaptation of the basic C-3 design which finally

[17] EVS (exposure-value system) was a simplified method of precomputed exposure which had a brief vogue as one of the transitional steps from manual to fully automatic lens adjustments which appeared in the late fifties. EVS was soon outmoded by the electric-eye cameras.

became the Autronic C-3 ($99.50).[18] A coupled rangefinder replaced the Argus 21's luminous-frame viewfinder, and the camera became the C-4 in 1951 and the C-44 in 1956.[19] In 1950 and 1951 the Model A dies were resurrected and a plug-in flash was added to make the FA. A squat black-plastic box made its debut in 1953 as the A4, priced at $39.95, including flash and case. When this model was phased out in 1956, it was replaced by the C-20, an almost identical camera with a restyled lens mount in a brown-plastic body to which a trouble-prone rangefinder had been added. When Argus designers of this period made a definite styling change, the results were far from original, aesthetic, or even interesting. Montgomery Ward bought the final C-20 production run in 1958, closing it out the next year in its photographic catalog at greatly reduced prices.

With its heavy reliance on the dependable C-3 (selling at $66.50 with case and flash) and the ability of Homer Hilton, vice-president in charge of sales,[20] Argus managed to main-

[18] When the 21's streamlined shape failed to outsell the sturdy boxlike C-3, the company designers decided that updating the C-3 was the answer. To an extent they were correct, but their attempts to simplify the camera in the Match-Matic models were doomed to failure, and later tries at automation were engineering disasters as well as design nightmares. The use of German shutters in the Autronic proved once and for all that no amount of additional precision was going to work. Dealers today still sell older C-3's on a regular basis, but the jazzy models are virtually worthless. Both the Standard C-3 and the Match-Matic were manufactured from 1958 to 1966. The C-33 appeared in 1959 and died in 1960, to be replaced the same year by the Autronic C-3, which lasted until 1962.

[19] Both the C-4 and the C-44 died in 1957. The C-4R, incorporating a rapid film advance, was produced in 1958 only; the C-44R appeared in 1958 and lasted until 1960. Both series were equipped with an *f*/2.8 Cintagon lens, but the C-44R was also available with a six-element *f*/1.9 lens and a slip-on, slip-off coupled exposure meter.

[20] Homer Hilton was a promotional genius to whom salesmanship was an art, and he practiced it with an amazing degree of success. A native of Sterling, Illinois, Hilton was one of the original Argus executives and

tain its position behind Eastman Kodak as the country's second-largest manufacturer and distributor of photographic equipment into the mid-fifties, building total assets of over $9 million in the process. But changing times, increasing costs, and unpopular designs finally caught up with the company as the decade ended. Sylvania bought control in 1959, and three years later, in a transaction that still astounds some financial analysts, Sylvania transferred control to a small but profitable maker of inexpensive photographic accessories, Mansfield Industries. As one analyst wrote, it was almost as though Sylvania were willing to pay someone to remove the Argus albatross from its neck, for Mansfield absorbed the larger, loss-prone Argus on a long-term payment-out-of-earnings deal. Almost no money changed hands, and that proved to be one of the problems. While Mansfield had been highly successful in selling inexpensive movie editors and projectors and imported accessories, it was an entirely different matter to market an obsolete camera line faced with foreign competition and with almost no working capital.

Mansfield's management was simply not up to the task, and the company sank into the Argus morass. A new slide-projector line ran head on into the instantly successful

---

remained with the company until he retired in 1952. He was one of the few members of the photographic fraternity whose achievements made him well known within the industry in his own time and still remembered today. He was best described by one of his subordinates as a "genuine gentleman." William F. Armstrong, in a letter to the authors dated December 10, 1969, drew this portrait of Hilton:

"Homer was, without a doubt, one of the most striking and memorable individuals I have ever met. His appearance was almost 'theatrical,' reminding one very much of Lionel Barrymore, with a flowing head of gray hair and a booming voice that could be heard for miles. His sales meetings were virtually theatrical productions, with very little advance preparation. He had the ability, however, to speak extemporaneously in

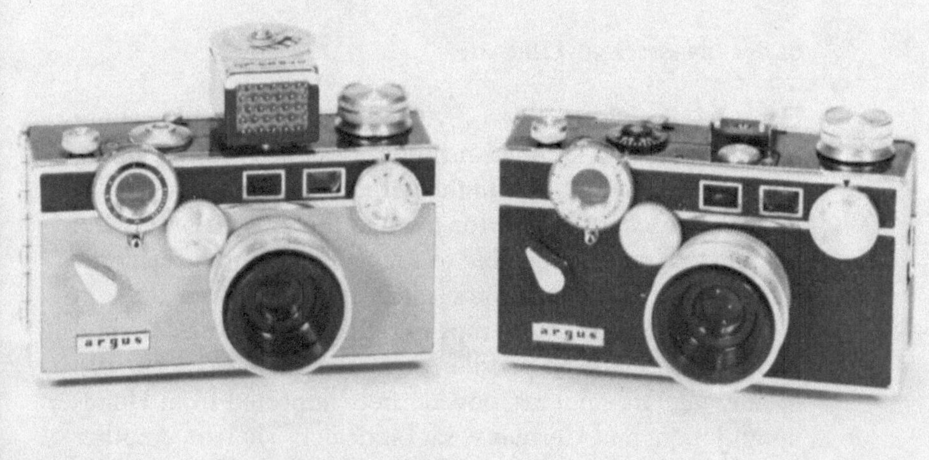

As the public became more sophisticated, the designers updated other features on the Standard C-3 (right) with a new cocking lever, a new shutter release, and, at last, a new lens mount which allowed the user to see the lens opening from the top. The Match-Matic C-3 (left) had a meter which read in EVS numbers and was the only means of setting the lens and shutter speeds accurately. Without the meter, which gave endless trouble, the camera could only be set by guess.

a manner I have seldom seen equaled. His personality was such as to instill the greatest respect of all of those working with him and to draw out that little extra bit of effort beyond the call of duty, 'just because Homer asked for it.'

"In the photographic industry, he was known as 'Mr. Photography' and was probably one of the best-known members of the industry, both nationally and internationally. A dedicated individual who worked extremely hard when he worked, Homer also knew how to relax and enjoyed himself when the work schedule was completed. He did not believe in an office intercom or sending a secretary with a message. The calm of the office was frequently distorted when from his office, a booming resonant voice would emanate with, 'Hey Bill, come in here a minute.' "

Hilton was undoubtedly one of the few genuine personalities who emerged in the research for this volume—and, according to his daughter, he never took a picture!

Carousel line introduced by Kodak and the top-selling Sawyer projectors. A beautifully designed 8mm-camera line was wiped out overnight with the advent of Kodak's Super 8mm concept and for a while the only products the company seemed to be able to sell were its 8mm projectors, which were made in a new plant that had been built in North Carolina in an attempt to reduce labor costs. Several efforts were made to peddle Argus as a good tax write-off. Meanwhile, its cameras, now entirely imported from Hongkong, Japan, and Germany, sat on dealers' shelves. Another new line of slide projectors designed outside the company failed after a year or so and was heavily discounted without much sales success. As the losses mounted into millions, Sylvania (which had publicly stated that it had no further interest in the company) was forced to resume ownership. Argus regained its corporate entity in 1967.

Shortly before the huge Expo 69 photographic show in New York City, it was announced that Italian interests had purchased the company, this time for cash. The Argus representatives at Expo 69 were confused by the turn of events but much more confident about the future than they had been in years. So far, however, in spite of a line of well-designed cameras using 126 film, made primarily by Balda in Germany, success has eluded the company. Argus reported a net operating loss of $2.75 million for the fiscal year 1969, approximately the same loss as that reported in the previous year. Fiscal 1970 continued in the red, with a further loss exceeding $1 million in the first three months. At the time of this writing, Argus was again being reshaped in the hope of finding another Charles Verschoor to lead the firm out of the wilderness. In an effort to regain a share of the market, a new line of 35mm single-lens reflex cameras, the Argus-

Cosina STL 1000 series, manufactured in Japan, was introduced (to the bewilderment of those who remember the company's unsuccessful attempt to market the Argus SLR a few years back). It is to be hoped that the new line will be successful. But the industry has changed greatly in the years since the end of World War II, and Argus must again learn to respond to the market with the same urgency and innovation it once had in abundance.

While the proud name of Argus has fallen on hard times, there is no denying its importance to the growth and popularity of the 35mm movement in America. The Argus A established the 35mm as a contender, the C-3 earned a place in the Smithsonian Institution for its unprecedented sales record, and Verschoor's vision brought forth a profusion of good, reliable, pacesetting designs. The C-3 was shaken on its throne in the early sixties by the increasing popularity of the 35mm single-lens reflex, and the Kodak Instamatic concept dealt it a death blow. The American photographic industry no longer even tries to compete in the world market at the price-and-quality range to which Argus addressed itself in the late thrities. The C-3 was one of the final holdouts among cameras which have passed into oblivion after having accomplished their purpose. The 35mm format is now engaged in a deadly struggle with the instant-loading cameras, and while no one seriously expects the 35mm to disappear, the future heavily favors the 126 format. It is quite likely that three decades from now the 35mm camera of today will be regarded as a quaint antique of a vanished era. But we would be the last to say that to the thousands of contented C-3 owners still clunking away.

10 . . . Followed Closely by . . .

While the International Research Corporation proved that a large and ready market existed for inexpensive 35mm cameras, it owed much of its success to two men who had preconditioned the public to accept less than the best if the price was right. Having already carved out a comfortable share of the market for inexpensive cameras, Otto Wolff Githens and Jacob J. Shapiro managed to make their fortune by placing millions of inexpensive cameras in the eager hands of people across the nation. Those who remember the Universal Camera Corporation are prone to a derisive reaction when its products are mentioned,[1] but if Argus popularized the 35mm concept, UniveX revolutionized the inexpensive-camera market,[2] and Universal produced probably the most innovative 35mm cameras in the under-$50 price range of the era. The story of Universal is an interesting one, as were its cameras, and we shall depart from our usual

[1] One former employee of Universal interviewed for this book turned blue with anger as he thought back thirty years and became so enraged that he was literally speechless. We never learned the reason for his unusual reaction.

[2] These very inexpensive cameras interested thousands of people in photography as a hobby, creating a wide market of potential 35mm users whose experience with the small negatives overcame the purists' objections to the negative size.

format in telling it, to include a brief description of some of its innovations outside the 35mm field.

When Githens and Shapiro went into the photographic business together in 1933, neither had much more than a passing acquaintance with the field, occasionally snapping a picture or two of their families or friends on weekends, as did millions of other Americans. The two had first met in 1918, when Githens, head of a New York loan company which was financing a taxicab operation in Indianapolis, had occasion to call upon Shapiro, vice-president of an insurance company which specialized in insurance for taxicab fleets. The two men kept their business acquaintance alive over the following years, and in late 1932, Githens learned that Shapiro was firmly convinced that the cheap camera was the answer to his personal financial depression.

Never one to turn his back on opportunity, Githens listened carefully and before long was actively contributing to a plan that would soon result in the Universal Camera Corporation. The two men concluded that the real profit was to be made in selling film and that the cheaper the film the more could be sold. The way to sell film cheaply was to sell it in short quantities to the customer at a price he could easily afford. While there is some indication that they approached Eastman Kodak Company to manufacture and package film for them under a private label, it was Gevaërt, the Belgium-based film manufacturer, which agreed to make and package their film.

Neither Githens nor Shapiro was an engineer, but both were past masters at persuasion, and the acquisition of their first camera design set the tenor for Universal's business practices over the next two decades, giving the newly created partnership a shady reputation within the industry which it never lived down—not that it tried very hard to do so. They went to Lockport, New York, and asked the Norton Company to design and manufacture a camera for them. Norton agreed and not only designed the camera but also built the dies. Before production began, Githens and Shapiro apparently had second thoughts, or at least a disagreement with Norton, for they returned to New York City, and the design (slightly altered in dimensions) appeared in October, 1933, under their own manufacture as the UniveX A.

Flabbergasted by this turn of events, Norton found itself with a set of dies and considerable money invested in the project. At Githens' request the camera had been designed with a unique film system: the winding knobs of the camera were the ends of the film spool. No film existed for this design, but Norton contacted nearby Kodak for help in an effort to recoup its investment. Eastman agreed to manufacture both the film and the spools and did so until the beginning of World War II.

With financing by Philip Maslansky, of the New York Merchandise Company, Universal's first camera was an unqualified success.[3] Incorporating an imported hand-polished lens and a simple but reliable spring shutter, the UniveX A sold for a retail price of 39 cents. Its most important market avenue was mass distribution through New York Merchandise as a premium or giveaway item. There was a method in this madness. The usual premium comes into the hands of its recipient by way of a third party, with no direct con-

[3] One of Universal's proudest achievements was the capture of a Los Angeles bank robber, who dropped a UniveX A while making good his escape. Police developed the film and found a recognizable photo of the well-known robber, who was taken into custody with the $10,000 he had stolen.

tact between consumer and product manufacturer. The UniveX altered the traditional pattern by using a special film spool which required the owner to purchase and use only the film marketed by Universal.[4] Although the price was ridiculously low by today's standards, the camera took pictures of good-enough quality to ensure its continued use by the owner, unlike other giveaways, which were usually in truth throwaways.

When its doors first opened, Universal's daily output was said to be 500 cameras, and within three years 3 million UniveX A's had reportedly been sold.[5] Production was steadily increased until 1935, when Universal set a sales goal of 4 million and claimed shipments of 20,000 units daily to department stores, drugstores, cigar stands, and camera shops across the nation.

In 1934 the Norton camera appeared, with Kodak-produced film. It was almost a duplicate of the UniveX in appearance but was slightly larger in body and viewfinder size and cost 50 cents. Since Universal had been first on the market with its version of the camera, Norton's attempt to recoup its investment was quickly denounced as a flagrant copy of a success. Few knew the whole story.

In 1935, Universal sold 1 million UniveX A cameras to a large drug concern and bought out the ailing Norton line, adding it to their own line as the Norton UniveX. Two less successful imitators were the Varsity and the ULCA. Even Woolworth got into the act with a 40-cent camera (20 cents

[4] Other premium cameras such as the Cub (later called the Clix and priced at $1.00) sold for as little as 15 cents but used standard film—usually 828 or 127—which was more expensive than the camera.

[5] Using a patented V-slotted spool flange, the OO film gave Universal a monopoly on the film business of its camera purchasers, and no one else was licensed to make it.

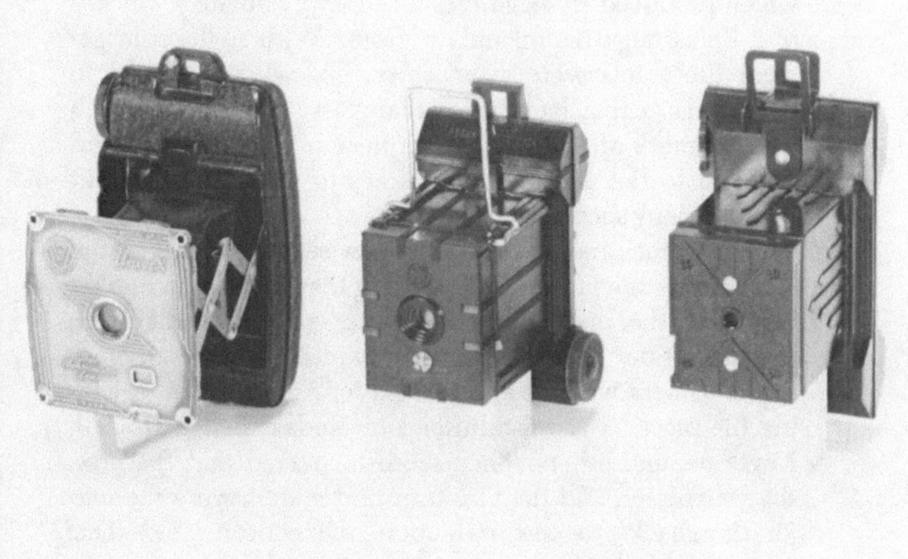

Three Universal models. Left to right: The small folding Deluxe AF-3, the original 39-cent UniveX A, and the Norton. Note how closely the slightly larger Norton compares with the UniveX A.

for the camera, 20 cents for the finder), but with its molded lens and rubber-band-operated shutter it quickly disappeared, along with several other minor makes. Kodak brought out its Bakelite 127 Baby Brownie for $1.00. Universal countered by introducing a line of three inexpensive folding cameras selling for $1.00, $1.50, and $2.50, and another million cameras passed into the consumers' hands within a year and a half of their introduction. In the process Universal had also sold 15 million rolls of its special OO film (for those who are interested, that is 2,800 miles of film), which produced six negatives a roll approximately the size of a Leica negative for only a dime.[6] With slight enlargement the prints were satisfactory, but when enlarged to post-card size they had the familiar glassy appearance which was the mark of an inexpensive piece of glass, ground and polished to the minimum necessary to call it a lens.[7] But who can deny success?

Not Githens and Shapiro, who now set their sights on the 8mm home-movie market. Realizing that they had a market which no other manufacturer of cameras had tapped as well, the two set out in the fall of 1935 to mass-produce an 8mm movie camera which could be sold for $10 or under, breaking the price barrier established by Kodak, Bell & Howell, Keystone, and others. They soon discovered that the lens, the viewfinder, and the film-transport system were the major drawbacks to cost reduction and concentrated their efforts on these three areas. Since the target was a mass market—and, more specifically, families that were interested in making their own movies but could not afford an outlay of $30 to $40 for a camera (to say nothing of the cost of a projector)—a theoretical quality level was predetermined. The ultimate goal was a movie camera, plain and neat in exterior appearance, functional to operate, and dependable in use, which would also take reasonably sharp and steady pictures. Ilex indicated that it could supply a lens

[6] The film-production figures given are those released by the company and are considered by many in the industry to be highly inflated. But whatever the true figures, Universal was selling an astounding number of cameras for a new company with an off-standard format.

[7] In 1936, Universal established a network of laboratories (in New York, Chicago, and Hollywood) and made a determined push through its retail outlets for the film-processing business, offering 3 × 4-inch prints from the postage-stamp negatives. Other photofinishers welcomed the move and were more than pleased to let Universal have the business.

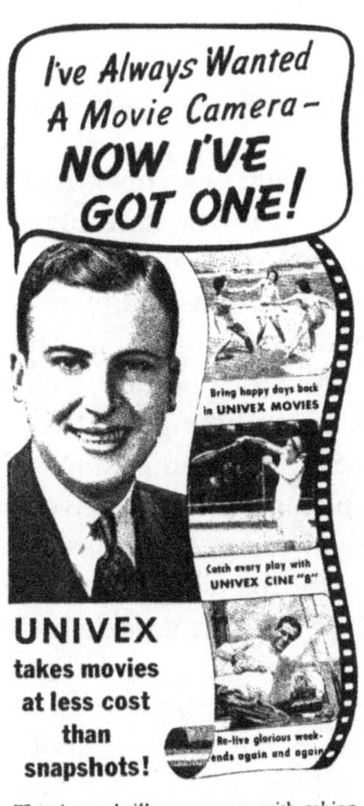

UNIVEX
takes movies
at less cost
than
snapshots!

There's no thrill to compare with taking your own movies. *Now you can afford it!* Here's the only precision-built, lifetime 8 mm. movie camera in the world under $30–the only one that uses inexpensive 69¢ UniveX Film. Compact . . . easy to load and operate. Stop in at your dealer's today and ask for free showing, or send for illustrated booklet. Universal Camera Corp., Dept. 68, New York City.

$9.95

WITH F 3.5
LENS $19.95

Prices higher outside U. S. A.

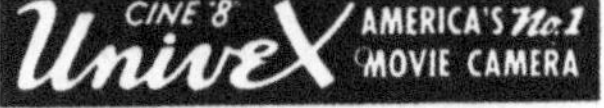

CINE '8' UniveX AMERICA'S No. 1 MOVIE CAMERA

NEW 1939
Movie Camera
FRONT-PAGE NEWS!

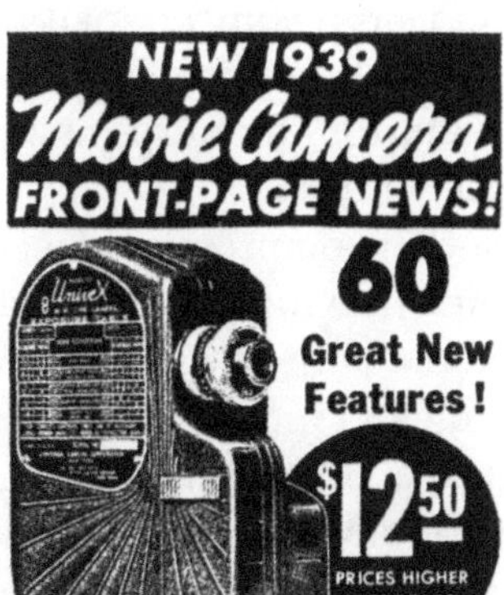

60
Great New
Features!

$12.50
PRICES HIGHER
OUTSIDE U.S.

Never before has a movie camera, at anywhere near this price, brought such perfection of detail . . . such simplicity of operation . . . such economy!

All America is talking about the new 1939 UniveX! Its magnificent new styling . . . its amazing performance . . . its 60 great features that you'd expect to find only in a high-priced camera! New built-in optical view-finder . . . new improved governor . . . new automatic shutter . . . new snap-lock hinged cover! Improved F: 5.6 lens, instantly interchangeable. New quiet, powerful, smooth-running spring motor.

INSURED MOVIES

An insurance policy backs this guarantee: *"Satisfactory movies if directions are followed or a new roll of film free!"*

CHECK THESE FEATURES

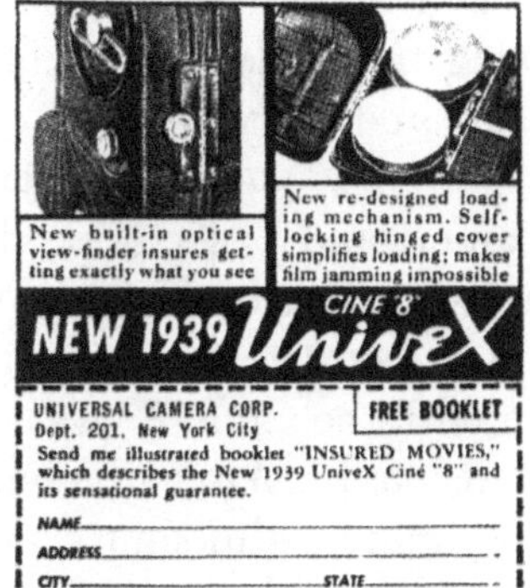

New built-in optical view-finder insures getting exactly what you see

New re-designed loading mechanism. Self-locking hinged cover simplifies loading; makes film jamming impossible

NEW 1939 CINE '8' UniveX

UNIVERSAL CAMERA CORP.
Dept. 201, New York City

FREE BOOKLET

Send me illustrated booklet "INSURED MOVIES," which describes the New 1939 UniveX Ciné "8" and its sensational guarantee.

NAME____________________

ADDRESS____________________

CITY__________ STATE__________

UniveX movie-camera advertisements of the 1930's.

adequate for the purpose at a reasonable cost based on a quantity order.

Another compromise came in the design of an open-frame viewfinder which folded out of the way for carrying (a telescopic finder was later offered as an accessory). The remaining and most important problem was the design of the film transport. While other manufacturers used a separate but carefully synchronized intermittent, or pull-down, claw and shutter, Universal succeeded in combining the two in a single unit which had a reciprocating action of approximately 1/30 second. With the problems resolved, Universal went into production, and its movie cameras reached the market in December, 1936, in time for the Christmas trade.

Housed in a die-cast all-metal charcoal-gray crackle body, equipped with an *f*/5.6 Ilex Univar lens, and priced at only $9.95, the original UniveX Cine "8" was considered an impractical purchase by most others in the industry, but it opened the door to home moviemaking, and in two years an eager public had bought 250,000 of the inexpensive cameras. Universal also supplied a companion P-8 projector utilizing the same reciprocating shutter-intermittent combination and carrying a $14.95 price tag. In that same two-year period 175,000 projectors were sold. Used as a team, the camera and projector produced acceptably sharp, steady pictures as envisioned. Their owners' satisfaction was a tremendous boost to the amateur cinema market. When film taken with the Cine "8" was shown with another projector, the projected image had a tendency to jump slightly owing to the difference in shutter-intermittent timing, but it was never really objectionable enough to hurt the camera's sales.[8]

True to their original concept, Githens and Shapiro were

still interested mainly in film sales, and the Cine "8" reflected their basic approach. It used special patented 30-foot spools of single-8mm film (as opposed to the 25-foot rolls of double-8mm film common to other movie cameras), which was supplied by Universal in two varieties: the Standard emulsion at 69 cents (in a red box) and Ultrapan at 95 cents (in a blue box). Processing was available for an additional 15 cents at the Universal labs in New York, Chicago, and Hollywood, and film sales kept all three labs operating at capacity. Single-8mm was briefly so popular that Bell & Howell, Keystone, and Revere adapted their models or designed new ones to accommodate it. The user was required to run the single-8mm through the camera only once. Double-8mm was run through the camera, after which the spools were turned over and run through a second time. The two sets of images were then split into single ones after processing and spliced together for return to the customer.

No cameras of other manufacturers would accept the UniveX spool, and records indicate that Ansco was the only manufacturer of the film. The bloom was off the rose by 1936, when Eastman Kodak introduced double-8mm Kodachrome. Sale of 8mm film spurted ahead, but, except for UniveX, it was almost all color.

Competitive single-8mm cameras died immediately. The closest competitor to the UniveX Cine "8" was Keystone's K-8, which featured the luxury of three running speeds, had

[8] Nor was the camera's frequent jamming owing to a weak take-up clutch, which seems to have been the major flaw in the Cine "8." Cine "8" cameras are relatively scarce today for two reasons. Owners had trouble obtaining film during World War II and gradually discarded the cameras. Trade-ins of the Cine "8" usually brought the owner $5 toward a more expensive camera, and once the deal was struck the camera-store clerk invariably took it from the customer and quickly deposited it in the nearest trash can, removing it from circulation for all time.

an *f*/3.5 lens, used both single and double-8mm film, and was made of stamped rather than die-cast metal. The K-8 was eventually dropped in price from $39.95 to $27.95. But ease of operation and low operating costs were the keys to the success of the Cine "8" system. It featured interchangeable lenses, allowing the use of a telephoto lens supplied by Universal or, with an adapter, any standard D-mount lens. While the camera could be purchased with any of five lenses, ranging in speed from *f*/5.6 to *f*/1.9, the *f*/5.6 lens kept the camera's price below the magic $10.00 mark and proved to be the most popular lens.

The year 1939, the year of the New York World's Fair, seemed an appropriate occasion for Universal to seize the initiative once again. Redesign of the Cine "8" had been undertaken once the original concept had proved salesworthy, and in December, 1938, the Exposition, or World's Fair, Cine "8" model was announced in time for Christmas sales and in anticipation of the opening of the fair the following spring. Thousands of Americans from all over the country would journey to New York City for the event, and many invested in the new camera to capture their trip on film.

Basically, the Exposition model was the same as the first Cine "8," but several refinements made it a much more desirable camera from the user's viewpoint.[9] The open-frame finder was replaced with an optical one built into the camera

[9] The Exposition Cine "8" was followed by the Cinemaster (prewar) and the Cinemaster II (postwar) models, using both single- and double-8mm film. They were much alike in appearance and resembled the Cine "8." But the Cinemaster, an engineering nightmare and an operational impossibility, was not understood by customers, and even the engineers could not make it work properly. Eventually the Cinemaster II was manufactured to use just double-8mm film, but even though it worked better than its predecessor, by that time the damage had been done to sales.

body, and a self-locking hinged cover replaced the removable twist-turn locking cover of the older model, simplifying film loading and reducing film jamming. The shutter was not self-capping, closing itself whenever the motor stopped and doing away with the annoying overexposed frames at the end of each run. The spring-wound motor featured a newly designed governor, running longer (up to 6 feet) and more quietly at a constant speed, with no slowdown as the spring tension unwound. Equipped with optical choices similar to those of the previous Cine "8," the antique-bronze Exposition model was introduced from $12.50 to $47.25 with the *f*/1.9 lens. Production of the older model ceased.

This brief account of Universal's early success is illustrative of the raw force and talent (and occasionally the use of questionable business ethics) with which American industry tackled the business of making photography a mass hobby and the small camera a genuine consumer item. Other manufacturers were to entertain success in this endeavor, but none achieved it with the motivations of the Argus and Universal companies. The Argus and its accessory equipment were primarily designed to sell cameras. Universal cameras were mainly a by-product of a new method of film merchandising at low cost and high volume, but Universal exhibited an amazing degree of inventiveness in adopting shortcuts which worked, and the inexpensive cameras were as much a part of the company's success as anything else. Many admired Universal for its brash audacity in producing and marketing what professionals knowledgeable about cameras and design principles tended to term "chrome-plated junk."

Since Universal maintained no important manufacturing facilities at the time, its cameras, projectors, and enlargers were composed of parts manufactured by other companies and assembled by semiskilled workers in Universal's loft factory at 28 West Twenty-third Street in New York City. Githens and Shapiro fully guaranteed their cameras against defective workmanship, and during this period they were so eager to build good will and keep the film customer happy that even if the camera had been run over by an army tank they would have gladly replaced it (and did so on many occasions) without question.

Having conquered 8mm home movies in 1936, Githens and Shapiro turned their attention to the rapidly growing 35mm field with the same determination: to produce a special 35mm film load and design a camera whose sale would assure them that the purchaser could buy film from no other source. In November, 1938, the instrument by which this market penetration was to be accomplished, the UniveX Mercury, was introduced. While other UniveX cameras had been designed around the OO spool, this one used its own special roll of film (not compatible with the OO). Once again the two men had returned to their original premise: the way to sell film profitably was to sell large quantities of short lengths for a low unit price. The Mercury's picture area was reduced from the standard double-, or full-frame, size (1 × 1½ inches) to single-, or half-frame, size (21/32 × ¾ inch). The film was packaged in rolls of either 18- or 36-exposure lengths and ranged in price from 30 to 50 cents, depending upon which length and which of the three emulsions the owner purchased.[10] Thus it was now

[10] For a brief time Universal also offered Dufaycolor film. This English product used a matrix system which, among other problems, made the film extremely slow in speed and in projection showed an objectionable pattern over the picture. Introduced before Kodachrome, it quickly declined after 1935.

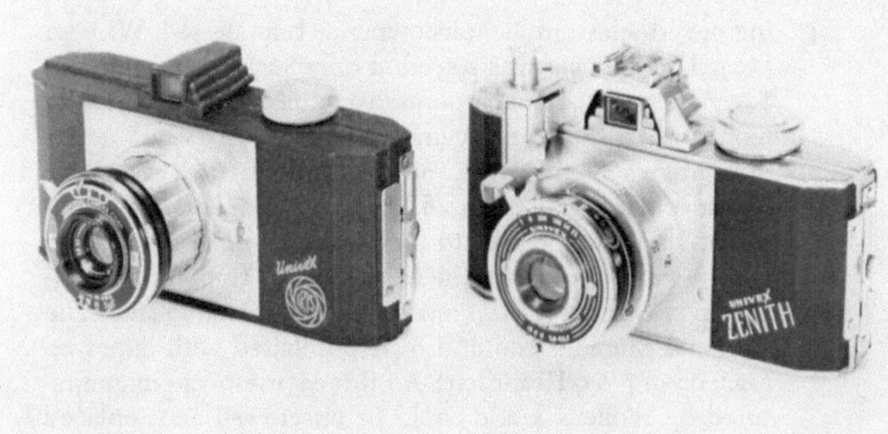

The Universal line included both the Iris and the Zenith models in several shutter and lens models. Both took OO film. The Iris body was die-cast alloy (pot metal), while the Zenith was made of aluminum.

possible to buy a camera which would produce 36 exposures for only 40 cents a roll.

The UniveX Mercury used standard 35mm sprocketed film attached to a metal spool with a small brass gear on the bottom flange. A paper backing provided protection from light fogging during loading and unloading, and no rewinding was necessary; the film was run through the camera and removed, and the empty feed spool became the new take-up spool for the next roll, doing away with the task of loading cassettes and magazines. In theory Universal had done it again with what was essentially a roll-film camera that took pictures in a variation of the 35mm format.

The UniveX Mercury, or Mercury I, as it is sometimes called, was a rather ungainly-appearing camera whose styling nevertheless grew on its owner as time passed. With an *f*/3.5 Tricor lens it was priced at only $25.00 (increased to $29.75 in 1940) and incorporated a number of features commonplace in the 35mm cameras of later years but rather innovative at the time.[11] Centrally located on the face, the Mercury's "instrument-panel" controls were functionally designed to allow the user to focus and set the lens, shutter, and frame counter from the front without tipping the camera up or turning it from top to bottom or side to side. The all-metal Mercury could also be purchased with either an *f*/2.7 or an *f*/2.0 Hexar lens. All three lenses were manufactured by Wollensak and could be unscrewed and replaced with the telephoto lens especially designed for the Mercury. Within its protruding forehead was a rotary focal-plane shutter which sliced light into chosen increments from 1/20 to 1/1000 second.[12]

Simple in design and engineering, the Mercury shutter delivered a comfortable whirring sound with each light-chopping motion, ending with a solid clunk as the cycle was completed, thereby assuring its owner that it had performed its function. The revolving shutter-selector knob was proof-positive to the doubter. As the exposure was made, the knob spun with the shutter, coming to rest in a position between two shutter speeds, visual confirmation that the shutter had indeed opened and closed. Accuracy? Actually it was quite good. Universal advertised that an observatory had testified to the shutter's accuracy, and it was sufficiently precise even

[11] Although Kodak and others had little use for Universal and its trade practices, there was great respect for the Mercury and a conviction that, if it had been designed in a full-frame format it could have been a tremendous competitor in the 35mm field.

[12] Its top speed was extended to 1/1500 beginning in mid-1939, and this model was priced at $42.50 with an *f*/2.7 Hexar lens.

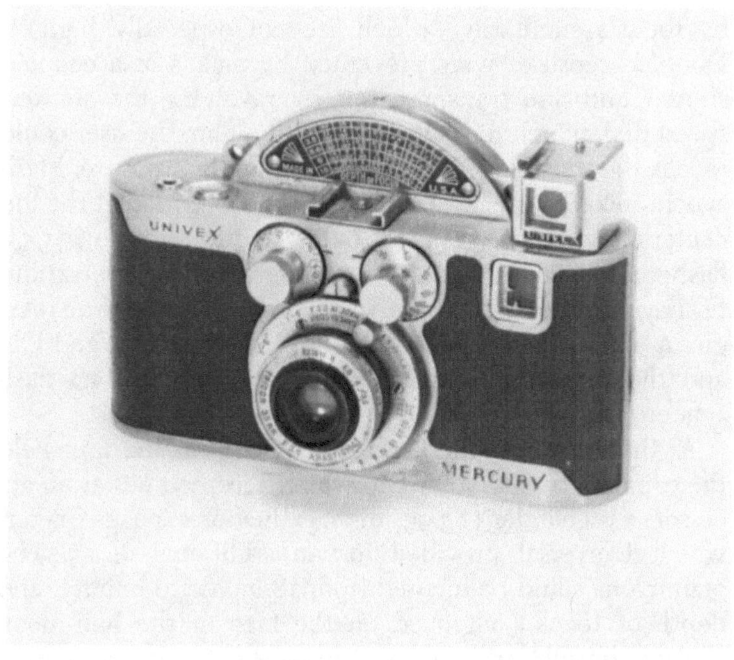

The flagship of the prewar Universal 35mm line was its single-frame Mercury, sometimes referred to as the Mercury I. Using a special film spool with paper backing to allow daylight loading, it had a unique and fairly reliable rotary focal-plane shutter. The first models had speeds to 1/1000 second, but for a short time in 1939 a version was also marketed with 1/1500 as the top speed. This early model has the accessory extinction exposure meter which was used with the complex dial on the camera back. The meter became a standard feature on many Universal cameras. It consisted of a circular numbered wedge for exposure calculation. Early models of the Mercury appear to have been better finished and had better mechanical operation than the postwar models. They were also smaller, since the special film spool took less space than standard 35mm cartridges.

by today's standards (which are not especially high).[13] Double exposures were prevented by means of a coupled shutter and film transport, but by revolving the shutter-speed dial, which did not transport the film, the user could recock the shutter for an intentional double exposure. Flash synchronization by means of a hot shoe mounted at the top center allowed the use of FP bulbs, and for a mere $3.95 a flashgun was available to extend the Mercury's versatility far beyond the range of the very popular (and plastic) Argus A and its more comparable kin, the Argus C and C-2 and the Kodak 35's, which had self-timers but no flash synchronization.

At the far left of the flash attachment (as the user held the camera) was another clip which accepted either an accessory rangefinder ($5.95) or an extinction exposure meter, which Universal provided for an additional $2.75. The 35mm lens could be focused from 18 inches to infinity, and depth of focus (engraved on the face of the half-moon shutter housing) *was* sufficient in most cases to compensate for the occasional slight errors every amateur made in judging distances. If the camera was equipped with the *f*/2.7 or *f*/2.0 lens, another plate attached to the rear of the shutter housing provided the necessary information for the additional *f*-stops.

Directly below the accessory clip was the reduction-image

[13] According to Bud Fowler, the chief instructor of the Technical Training Division, National Camera Repair School: "The rotary shutter was simple in design, and fast shutter speeds could be achieved with relative ease. The speed-control mechanism was rather crude by current standards. However, the durability of the shutter itself was reasonably good, and the Mercury was easy to service in comparison to most model cameras." Interestingly enough, the shutter principle was reincarnated for use in current models of the Olympus Pen, made by Olympus Optical Company, of Japan.

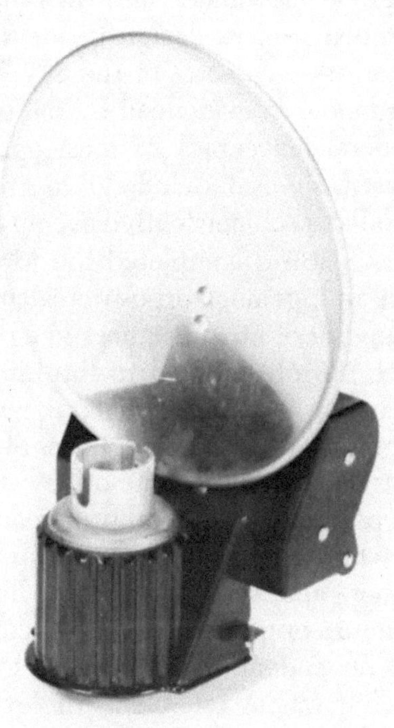

The only unique characteristic of Universal's flashgun was that it could be used with all the manufacturer's cameras except the Uniflash, the Steré-All, and the Minute 16. Even the folding cameras utilized this gun with an accessory shoe. The penlight batteries rode behind the gun, and the reflector was removable. The bayonet-bulb adapter in place was an afterthought by its owner—the unit as sold used only screw-base bulbs.

optical viewfinder, which rather inaccurately indicated what the lens saw. That Universal would incorporate such a parallax-prone viewfinder into an otherwise interesting and enlightened camera design is not so surprising when one evaluates other cameras of the day. Viewfinder design was in its infancy, as evidenced by the open-frame finders and the general inaccuracy of most optical finders being used. At least Universal included two small index marks to indicate parallax problems with close-up pictures.

Two other features completed the Mercury's functional innovations: an ingenious exposure calculator designed for just about any set of circumstances imaginable and a rapid film advance. Firmly secured to the hinged back, the exposure calculator was standard equipment, and the user had but to pose a problem to it in terms of film, lighting condition, shutter speed, filter, time of day, season of the year, and camera position, and the exposure could be determined. Of course, while the user was performing this interesting word, number, and symbol exercise, he could be losing dozens of picture opportunities. The accuracy of the device was entirely dependent upon the eye of the user, but with a little practice it was as reliable as the extinction meters in use at the time.

The rapid film advance was a $2.50 accessory which sat over the knurled film-wind knob and was secured to the cable-release socket with a screw. Once it was in position, a couple of quick, short, downward thrusts of the index finger wound the film and set the shutter. The user could expose an entire roll without moving the camera away from his eye, providing the exposure and focus remained the same on all shots. Although popular today, the rapid-advance feature was fairly unique with the Mercury at the time. Leica had an expensive special base with a hand-operated trigger and later a spring motor, and several attempts were made to devise rapid-advance levers for the Argus, but until the Kodak Ektra appeared in 1941 with a rapid-wind lever built into the interchangeable back, the Mercury system was the lone success among low-priced cameras.

While the Mercury performed faithfully for its owner, it did have one major problem. The most common cause of breakdown was a design weakness in the film-transport and shutter-winding system. The film was advanced by the take-up spool, which, synchronized with the sprocket that wound the shutter, monitored the length of film being moved by the number of film perforations and counted the exposures. All these operations had to take place in absolute harmony. If not, the usual result was that the film take-up would begin dragging the film past the sprocket, ripping out the film perforations. It was not too unusual for a Mercury owner to open his camera and find three pieces of film, two ends that had once been the edges and the center containing the exposed pictures—and a large pile of film chips from the ripped perforations. (The Mercury II, introduced after World War II, also had this flaw, probably owing to hasty assembly to get the cameras to market.)

Weighing in at 18 ounces unloaded, easily held and used, the Mercury was a trim package which lived up to most of its promises most of the time. It drew quizzical glances from the uninitiated, but Mercury owners were fond of their odd-looking cameras and were especially pleased when Universal announced the Mercury II after the war ended, even though it was priced at more than triple the cost of the original Mercury.

Universal's second line was its Corsair, introduced in

1941 at $16.75. The original Corsair was also designed for the special UniveX film, but the Corsair II, which used a standard 35mm cartridge, followed shortly at $19.75. The Corsair also had an interesting design. The body was molded of plastic. The lens mount, winding knobs, tripod socket, frame counter, and accessory clip were metal. It featured a noninterchangeable three-element *f*/4.5 anastigmat lens in a rim-set leaf shutter with speeds of 1/25 to 1/200, *T* and *B*. The lens mount was a collapsible tube similar to that of the Argus A, but was locked and unlocked by revolving the focusing scale about 1 inch beyond the infinity setting instead of by bayoneting. The shutter could not be tripped when the lens mount was retracted, and flash synch was incorporated by means of a hot shoe in the accessory clip. The Corsair's dies were somewhat crude, resulting in a not-too-well-finished product, but it was produced when Universal's fortunes were reaching a low point; the war had cut off foreign film supplies, and a camera line taking standard film cartridges had become essential. Its outstanding feature was a natural feel in the user's hands. Fairly thin and somewhat resembling the Kodak 35 camera, the Corsair rested easily in the hand, unlike some of the more awkwardly designed 35mm's of the period.

The basic Corsair dies became the postwar Buccaneer, introduced late in 1945. In the Buccaneer the plastic viewfinder–extinction-meter unit[14] was replaced with a chrome-plated metal housing extending across the camera top. The

[14] This was one die from which the company got much mileage. It appeared on the postwar Meteor, a 620 roll-film box camera using the lens and shutter of the Roamer I folding camera and the Mercury flash unit. The first Meteor (followed by a Meteor II) was a classic example of a camera built from excess parts; while it took fair pictures, it was one of the most ungainly and awkward cameras Universal ever produced.

housing contained a coupled rangefinder with its superimposed image centered in a small spot in the viewfinder. The old reliable extinction meter was placed beside it on the right. The lens was a coated *f*/3.5, and the shutter range extended from 1/10 to 1/300, *T* and *B*. The Buccaneer sold for $65.00 (plus case, $7.25), but it ran a poor second in popularity and sales to the Mercury II, which also made its appearance late in 1945. The Buccaneer's problems lay in the design of its rangefinder, which was impossible to keep in adjustment. When the Buccaneer was finally retired, the old viewfinder–extinction-meter combination was resurrected, and the Corsair-Buccaneer body reappeared as the Vitar. Never advertised, the Vitar was sold as a promotional item, usually at 50 per cent below its inflated "list" price.

Basically the same camera as the original Mercury, the Mercury II was slightly enlarged in design around the standard 35mm cartridge, and while it lacked that indefinable aura which had made the Mercury I attractive to so many buyers, its ability to use color film (even if processors did not mount it) enhanced its sales considerably.[15] The top shutter speed was now 1/1000 second, reduced from the 1/1500 on some models of the Mercury I when Universal found that it failed to perform at its rated speed because of an inability to maintain proper spring tension after repeated use and also contributed to more frequent shutter trouble. All in all, the Mercury II was a natural evolution in design (a

[15] In an effort to overcome the handicap posed by wartime unavailability of its special film load, in the spring of 1940 Universal provided a daylight loading cartridge of its own design for those who wished to use bulk film with the Mercury I. The loader enabled the crafty cameraman to use Kodachrome in his Mercury by unloading the Kodak cartridge into the UniveX loader and then replacing the film in the Kodak cartridge before sending it to Kodak for processing.

characteristic of Universal products) which had the same flaw as its forebear, a film-transport–shutter-wind problem.

With its original styling the Mercury II was contemporary in design, perhaps a bit too much so. Another flaw lay in the finish, which seemed to grow worse as the firm's era drew to a close. Apparently the aluminum body was covered with flashed chrome, which lasted until the proud customer got home with his new purchase—if he was lucky. Shortly before the war drew to a close, Universal, along

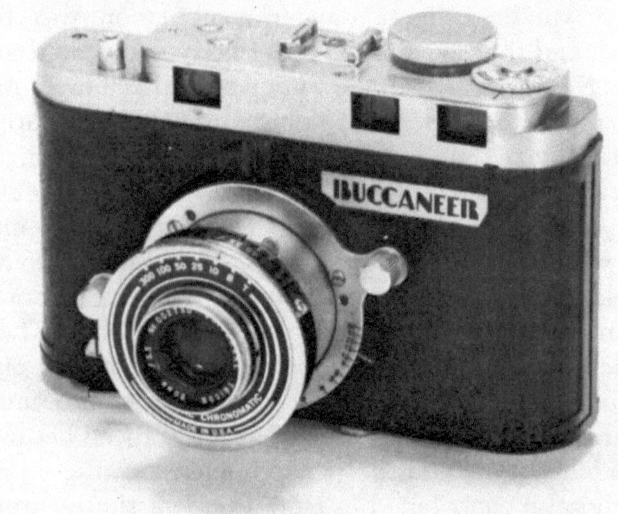

The Buccaneer (left) was a unique camera and probably one of the most profitable models Universal made. It began life as the Corsair before World War II, sporting a combination viewfinder and exposure meter that was to see service in many other Universal models. The Corsair came in two versions: the Corsair I (right) which used special Universal loading spools, as did the Mercury; and the Corsair II, which used the standard 35mm

with Argus, hit a bonanza in the form of priority orders from the army post-exchange system. Both companies were given huge orders of cameras to stock PX shelves and exchange for some of the dollars in the hot hands of GI's overseas. The army PX service placed orders totaling over $1 million with various camera manufacturers even before a single camera had been produced. Rumors in the industry said that Argus' share was 50,000 C-3's, which did much to make that camera the most popular 35mm in history.

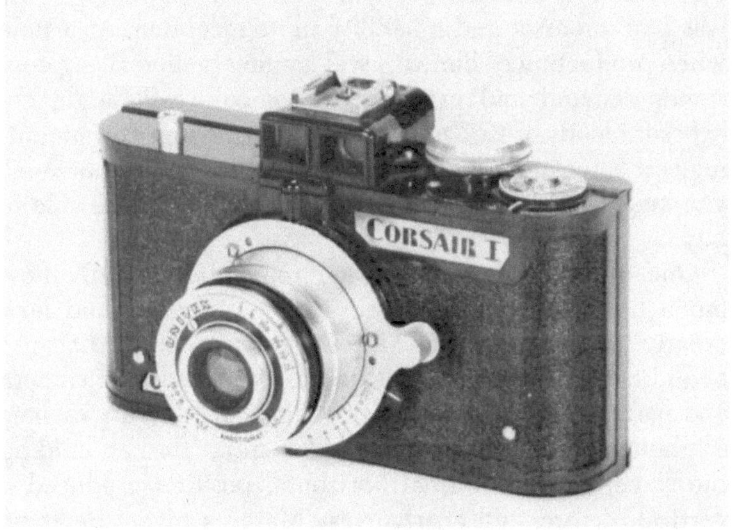

cartridge. The Buccaneer was equipped with a rangefinder that gave constant trouble. Finally the rangefinder was removed, and the camera was renamed the Vitar. The Vitar, the old Corsair with a new nameplate, was one of the last cameras produced by Universal. The Buccaneer's most interesting feature was that its designer had devoted himself to making easy operations as complicated as possible.

Universal's share is unknown. It is generally believed that the half-frame size and the absence of a rangefinder on a camera priced above the C-3 made it unpopular with servicemen. In spite of its *f*/2.7 lens and trimmer shape, the Mercury was almost always available in a PX, while the C-3's were snapped up as soon as they arrived. But the Mercury II was popular on the civilian market in those days, if for no other reason than its availability. It was sturdy and usually dependable, and it listed at $82.90 (plus case, $7.25). In 1946, the film economy of color-slide photography was both an asset and a liability in its reception. At a time when production of film was still lagging behind rising consumer demand and prices were comparatively high, 20-exposure loads of Kodachrome with processing and mounting cost $4.00 and up, depending upon the dealer, how well you knew him, and how many boxes of film he was able to buy.

One of the authors (Lahue) remembers vividly how much he admired the Mercury as a youngster and how greatly he despised most Mercury owners, partly because he did not have the good fortune to own his dream camera and partly because he resented failings of Mercury owners as photographers. Since it had a half-frame format, holding the camera in the normal horizontal position produced a vertical picture, and nearly every Mercury owner he knew took pictures that way, regardless of the rules of composition.

Kodak did not mount single-frame color transparencies at the time. That bothered Mercury owners not a whit; they apparently enjoyed showing their pictures in filmstrip form, and if and when they had the time, those frames worth keeping could always be mounted in standard 2 × 2 Jiffy

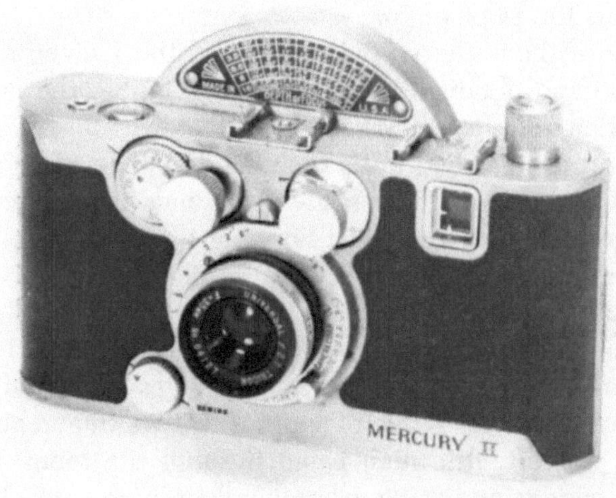

The postwar Mercury II was to be the ultimate in film economy and quality—or so the ads promised. But owners found that their money had bought more problems than pictures; there were frequent difficulties with the film transport. Universal had apparently realized that the public wanted color, and since Kodachrome came only in a standard cartridge, the body of the Mercury was enlarged to accept it. Use of the standard cartridge also made a means of rewinding the film necessary, a factor which led to the transport problems. In other respects the camera remained much the same as the prewar model. The f/2.7 lens produced by Universal was standard, but f/3.5 and f/2.0 models were also available. While Kodak provided Kodachrome, the company would not mount it in the single-, or half-frame, format, and this, along with the camera's mechanical problems and its poor finish combined to eliminate the Mercury II after a few years. But in its unusual appearance the Mercury series had few peers.

(paper) mounts supplied by Universal (and others) at 55 cents for a box of 50.

One of the authors (Bailey) visited the Universal plant in 1948 and obtained interesting insights into the postwar operations of the company. The plant was on an upper-floor loft and seemed to be tucked into nooks and crannies which had originally been used for some other enterprise. It was difficult to guess just how big the plant was, but in a letter to the author Universal claimed that it utilized 125,000 square feet of rented floor space. Its bankruptcy petition claimed 70,000 square feet. Boxes of parts littered the halls, and products in various stages of manufacture were everywhere. While the author was waiting in an anteroom, a short little man came through the room several times, coming and going to another floor by elevator. Finally the author engaged him in a short conversation and asked how Universal was managing to stay in business.

"Devega!" he replied.

"Devega? What's that?"

"Devega Sporting Goods!" and he departed in the elevator for the last time.

It developed that Devega Sporting Goods, a chain which eventually went bankrupt itself, was selling Universal cameras at half the list price. Universal's profit margin must have been high, because the company was apparently staying alive by making the products for sale at that price.

Finally the author was taken on a fast tour of the plant. Like many factories it was rather dingy. In the rear a lone woman was working at a machine engraving the lettering in lens mounts and filling the engraving with white paint. Next door to her was the lens-grinding shop. The author did not know enough about lens grinding at the time to understand much of the proceedings, but it was obvious that it was a do-it-yourself operation. There were rows of drill-press heads set at angles for the polishing heads for the lens banks. It all looked very crude and dirty—but some aspects of lens grinding are dirty and rather crude. Throughout the plant lay dozens of finished castings for projectors. The castings arrived at the plant finished, painted, and ready for assembly. It was explained that Universal did no casting (and, it is assumed, little machining). Universal was primarily an assembler, which was not unusual among small-camera manufacturers.

In another room were thousands of Minute 16 cameras on a huge table. The guide said that Universal had over $1 million worth of orders for these cameras and that they marked a new era in photography. But it was an obvious last-gasp effort to survive and also a final attempt to establish a market for another film size and a special spool. While it had the advantage of cartridge loading, the camera arrived on the scene shortly after the Whittaker Micro 16 and many Japanese imports which also used the small-size film, and no respectable network of processing stations could be established in time to save the Minute 16 from failure.[16]

Almost everything Universal produced was die-cast

16 The Minute 16 appeared near the height of a subminiature camera boom to compete with the Micro 16, the Tynar (both American-made), the Minox (German-made), the Gami (Italian-made), and a host of Japanese models. The absence of an adequate film supply in a standard cartridge, along with mediocre quality at best (the 16's film transport frequently jammed, a problem found to some degree in virtually every other Universal camera design not using OO film), killed it quickly. Photofinishers shuddered during this period, for processing was becoming fully automatic, and the investment required to process subminiature film was far too large for what they felt to be meager returns. Universal converted its three labs to handle a rush of Minute 16 business which never arrived, and the unreturned cost did not help the company's balance sheet.

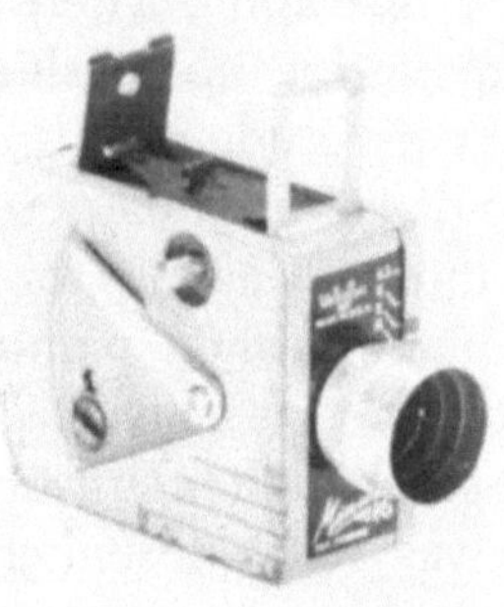

The Minute 16, Universal's contribution to the ill-fated 16mm boomlet of the early fifties wrote finis to the history of the manufacturer. A mediocre lens imported from Germany was coupled by slipshod workmanship to an erratic shutter. Yet Universal reported more than $1 million worth of orders for the camera. Several changes were made in the inner mechanism, but the only outward design change was in the viewfinder—models opened and shut in a manner suggesting rabbit ears. The f/6.3 lens was hardly a spy's delight, but many stores advertised such cameras as genuine spy equipment. Fortunately for spies, the war was over, and the demand for such items was negligible in the trade. Perhaps to encourage less clandestine use, the camera was equipped with a flash contact built into its tripod socket, and the flashgun was several times the size of the camera. One Minute 16 rests at the bottom of the Pacific Ocean, where one of the authors (Lahue) deposited it after spending several days in an unsuccessful attempt to transport a cartridge of film through it.

(rather than stamped, which is usually a mark of superior and more expensive construction), resulting in a lightweight camera. That apparently hurt Universal's sales. There was not enough heft to the cameras; they lacked the "feel" expected of good cameras. While the lenses were manufactured by standard companies like Ilex and (before the war) Wollensak, the front shutters were not. Though it seems improbable that Universal would go to the expense of tooling for a shutter, even the rather simple one Universal used, apparently the company did just that.

The Universal Camera Corporation closed its doors in 1952, two decades after they were opened. Many factors contributed to its business failure. When Germany began its march across Europe in September, 1939, Universal's supply of Gevaert film was then temporarily shut off.[17] Gevaert's American distributor acquired a factory in Williamstown, Massachusetts, in late 1939, and began manufacturing sensitized products in 1940. But with American entry into the war in December, 1941, many products became difficult to obtain. The OO film was soon in short supply. The Mercury cartridge mentioned earlier allowed its owners to continue using their cameras, but there was no substitute for OO film.

Because of the shortage of film, by the close of 1940, Universal was nearly bankrupt. But two years later, in one of the most remarkable business turnarounds in American industrial history, it had net sales in excess of $5 million and was employing 1,100 people, winning an Army-Navy E Award for aiding the war effort. Business had never been better for Githens and Shapiro, who had obtained a contract to supply

17 Much of Gevaert's photographic paper stock, gelatin, sensitizing dyes, and other chemicals had been imported from Germany.

binoculars to the Allied forces. Under General Manager Percy Case, Universal turned from an operation employing semiskilled workers to one of automatic machinery and unskilled labor. Its binoculars were rated excellent, an outstanding achievement for a firm which had never before ground a lens.[18]

After the war ended, when Universal turned its attention back to the consumer market, it introduced a line of twin-lens reflex cameras (Uniflex I and II), three folding cameras (Roamer I, 6.3, and II), and sound and silent movie projectors in 8mm and 16mm, in addition to the Mercury II, the Buccaneer (later the Vitar), and the Cinemaster II series. Universal enjoyed a tremendous expansion in the immediate postwar boom, and employment rose to almost 1,700, with subcontracting of components dropping to 50 per cent (primarily screw-machine parts, die castings, and plastic moldings). Using its wartime experience in lens grinding, Universal now produced most of its lenses, claiming one of the largest and most progressive optical shops in the business.

But Universal was caught with huge inventories in the 1948–49 recession, which severely affected the photographic industry, and the company never recovered. The number of employees was drastically reduced until it stabilized at 190 on January 1, 1950. New York Merchandise Company now owned approximately 50 per cent of Universal, and after a loss on the books of over $3 million during 1947–50, bankruptcy proceedings were initiated on April 16, 1952. Yet during the period in which the giant loss was incurred, Uni-

[18] After the war the binoculars were marketed until the backlog not required by the government was exhausted. The 6 × 30 prismatic binoculars were priced at $85 each plus 20 per cent federal retail excise tax.

versal grossed over $12 million, and one suspects that the organization was not actually bankrupt except by virtue of bookkeeping procedures made possible by tax write-offs and other provisions.

While there are probably many other factors that were never made public by the three major stockholders (New York Merchandise Company, Githens, and Shapiro), it is reasonable to assume that Universal's $2 million investment in the Minute 16 appeared unrecoverable, and the company was cashed in before the situation became really serious. Universal's attempt to diversify by designing a complex phonograph record changer cost a small fortune which was also never recovered.

Before we draw the curtain on Universal, one other interesting innovation should be mentioned. During the mid-thirties, Joseph Pignone, a young mechanical-engineering editor for *Mechanics and Handicrafts* started work on a revolutionary design: a single-lens reflex camera which would incorporate a French shutter mechanism within its lens, a principle now used in Hasselblad and Contaflex cameras. Universal hired Pignone, took a license on his patent, and set him to work developing a prototype around a 2¼-inch format. When the camera shutter was tripped, the mirror which provided the viewfinder image swung up out of the way, allowing the light from the lens to pass directly to the film (as in modern single-lens reflex cameras). The sound of the movement was excessively loud and distracting. Universal put Pignone to work on a 35mm version, reasoning that the smaller mirror would move more quietly. Pignone did not complete the assignment, for war came, and research and development ceased. Universal shelved and never resurrected the one product which might have made its future a different story. But there was a happy ending for Joseph Pignone. Today he ranks high among design engineers and has accumulated about 65 patents. Among his most recent projects was an automated enlarger for the Charles Beseler Company, of East Orange, New Jersey.

And so we end the story of Universal. Over the years its products have acquired a bad reputation which they do not quite deserve. The prewar cameras and accessories were unusual designs with individuality which today's cameras lack (in common with today's automobiles). While innovations were the result more of cost-reduction and profit-maximization goals than of the altruistic "best product for the least money" concept, that does not change the fact that Universal tried to be different and succeeded. It may not have made the best cameras of the era, but the Corsair, Buccaneer, and Mercury models certainly were among the most interesting.

11 Eastman's Grand Design

That Eastman Kodak Company contributed some of the most honored, best designed, and most unusual of the American 35mm cameras (as well as its share at the other end of the scale) was to be expected from a company which had followed closely the precepts of its revered founder. Although in recent years a vocal minority of its customers have accused the industrial giant of sitting on its hands and spewing forth a plethora of high-priced and noncompetitive photographic goods, nothing could be further from the truth. Picking up the threads of our story in the early thirties, we shall soon see that freshness and originality were not the sole province of the smaller companies.

Kodak's first 35mm camera, and certainly one of the most famous 35mm cameras of all time, was the Kodak Retina. The camera was the result of Kodak's purchase of the Nagel Camera Company, of Stuttgart, Germany, which became Kodak A.G. Strictly speaking, therefore, this camera was not an American 35mm, but, like the Leica and Contax, its impact on the American market was great, and without the introduction of the Retina and the film cartridge designed for use with it, American efforts to produce low-priced 35mm's like the Argus A would have been futile.

Moderately priced at $52.50 with an *f*/3.5 Xenar anastigmat lens mounted in a rim-set eight-speed Compur shutter (1 to 1/300, *T* and *B*), the Kodak Retina was first introduced to the American market in December, 1934. Combining simplicity with mechanical and optical precision and performance, it set a new standard in folding cameras. In closed position the camera measured only 4¾ × 3 × 1¼ inches and had the feel of a fine instrument. Three months later the camera appeared with a Compur Rapid shutter (giving an additional shutter speed of 1/500) at $57.50. In 1936, 1937, and 1939 further slight alterations appeared as various improvements were incorporated in the original design: the additional choices of Kodak Ektar and Zeiss Tessar lenses, a built-in shutter release, a double-exposure-prevention device, an accessory shoe, and so on. The small changes are too numerous to mention here in great detail,[1] but the camera's popularity was sufficient to produce a younger brother in 1937: the Kodak Retina II camera. The new model combined the basic simplicity and the sturdy design of the original camera with a coupled rangefinder and a fast *f*/2.8 or *f*/2.0 Schneider Xenon lens.[2] The expanded versatility provided an impetus for owners of the Kodak Retina I to trade up the scale.

Success of the new camera was made possible by concurrent introduction of Kodak SS Pan and Panatomic film

[1] Readers interested in the development of the Kodak Retina camera models are referred to the following publications, which are devoted to the camera and its use: *The Retina Guide*, by W. D. Emanuel; *The Retina Manual*, by Edward S. Bomback; and *The Retina Way*, by O. R. Croy.

[2] Priced at $115 or $140 with the Schneider lenses. It was also available with an *f*/3.5 Ektar lens. It was finished in satin chromium and black enamel and featured a body shutter release and double-exposure prevention. The 1939 model combined the rangefinder and viewfinder in one window.

in the new daylight-loading cartridges, ending the darkroom difficulties of handloading cassettes, and its popularity was never in doubt after the advent of 35mm Kodachrome film in 1936. Finished in black morocco leather and enamel with nickeled knobs and dials (in 1936 chrome-trimmed models were introduced), the Kodak Retina camera featured an enclosed optical viewfinder, a depth-of-focus scale, and duplicate focus and *f*-stop scales easily viewed in either the horizontal or the vertical position, all neatly and conveniently placed on the compact die-cast aluminum-alloy body.

The design was so popular that in 1946 production began again on both models of the Kodak Retina, with flash synchronization and hard-coated lenses appearing in 1949. Continued improvements in design and handling resulted in the Kodak Retina IA camera in early 1951, followed by the IIA in 1952. The evolution of these compact folding 35mm's into cameras with rigid lens mounts in the mid-fifties was a hard blow for the purist, and gradually the classic Kodak Retina became the Kodak Retina Reflex, whose various models met with indifferent success before being retired in favor of the new Kodak Instamatic Reflex.[3]

Rumors attributed Kodak's decision to let Kodak A.G. build its first 35mm miniature camera[4] to a reluctance on Kodak's part to commit itself firmly to the new film size, believing that 35mm would never develop into a market

[3] The early models of the Kodak Retina Reflex camera allowed the user to alter focal length by replacing the front elements of the lens. Later models and the Kodak Instamatic Reflex camera allow for full lens interchangeability.

[4] It was a sound move; there was a market in Europe for such a camera, and the German styling was classic in its simplicity and clean lines. The Kodak Retina would have met with success even if it had never been introduced in the United States.

PHOTOGRAPHY'S NO. 1 GIFT THIS YEAR

KODAK RETINA II

A Superb New Miniature "Still" Camera

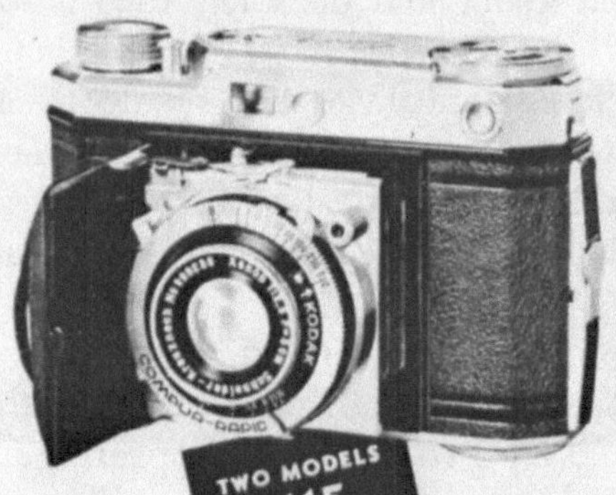

RETINA II is a truly great camera—a gem of precise mechanical excellence and operating proficiency—and it looks the part; a beautiful piece of camera craftsmanship.

Comes in two models—identical except for lens—anastigmat *f*.2.8 or *f*.2.0—$115 and $140, respectively, including tan leather case with neck strap. The camera case and back are of die-cast aluminum alloy, covered with black leather. Exposed metal parts are in satin-finish chromium and black enamel.

Housed in a metal turret are its direct view finder, coupled range finder, exposure-count dial, lever controlling clutch for forward and reverse winding, and brake lever for rewind knob. A body shutter release is provided, and a novel device prevents the user from making double exposures.

Three Supplementary Lenses

Three supplementary lenses may be had for close-up work; and a variety of filters. Retina II takes Kodak Panatomic and Super X Film, in 18- or 36-exposure magazines; Kodachrome Film, in 18-exposure magazines; Kodak "SS" Pan and Infra-Red Film, in 36-exposure magazines. All 35 mm.

See Kodak Retina II—at your Kodak dealer's. It's the last word in Eastman fine cameras—and first choice for a breath-taking photographic Christmas gift.

Only Eastman Makes the Kodak

EASTMAN KODAK COMPANY, Rochester, N. Y.

The German-made Kodak Retina II series represented the best that Eastman Kodak Company had to offer the American public. Above are shown the first II (advertisement, top), a postwar II (top, right), and the IIa (bottom, right), the last of the noninterchangeable-lens models. To many the IIa was the end of the classic line, although the IIc and IIIC models that followed were

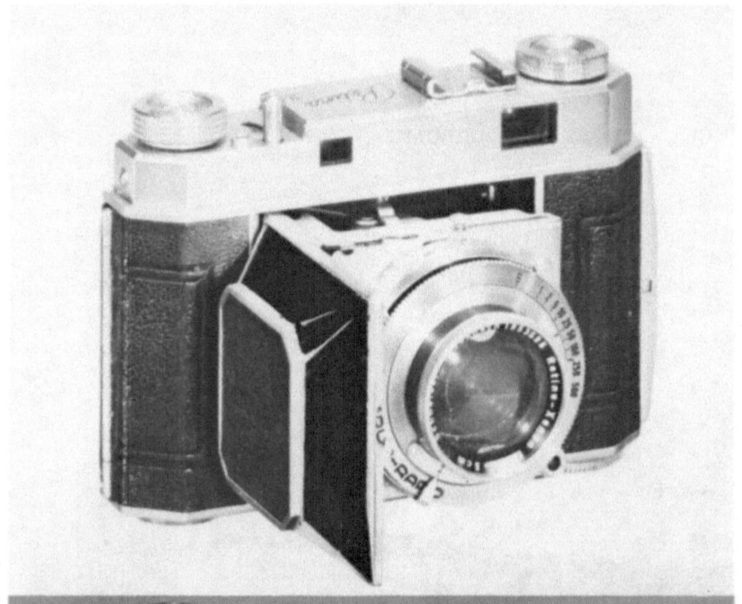

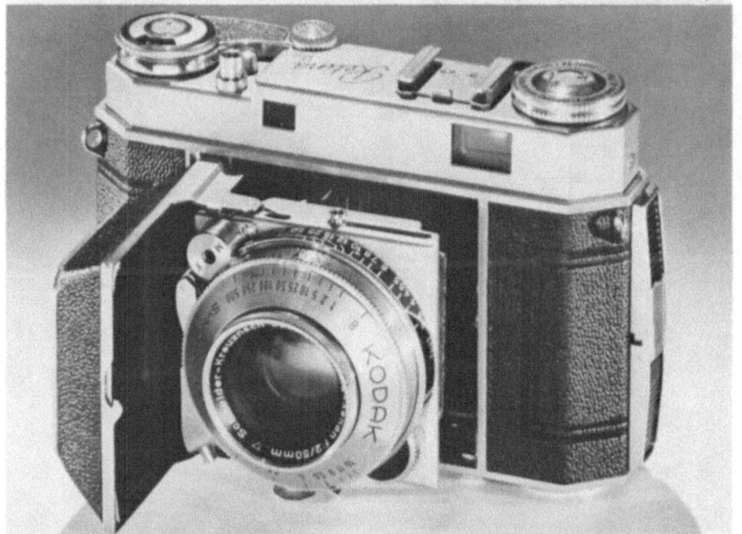

extremely popular and convenient, with their built-in meters and cross-coupled shutters. But the IIa was a compact handful with a sharp lens, rapid film wind, and a fully synchronized shutter. In the authors' opinion, only the lack of a bright-frame viewfinder kept it from being perfect.

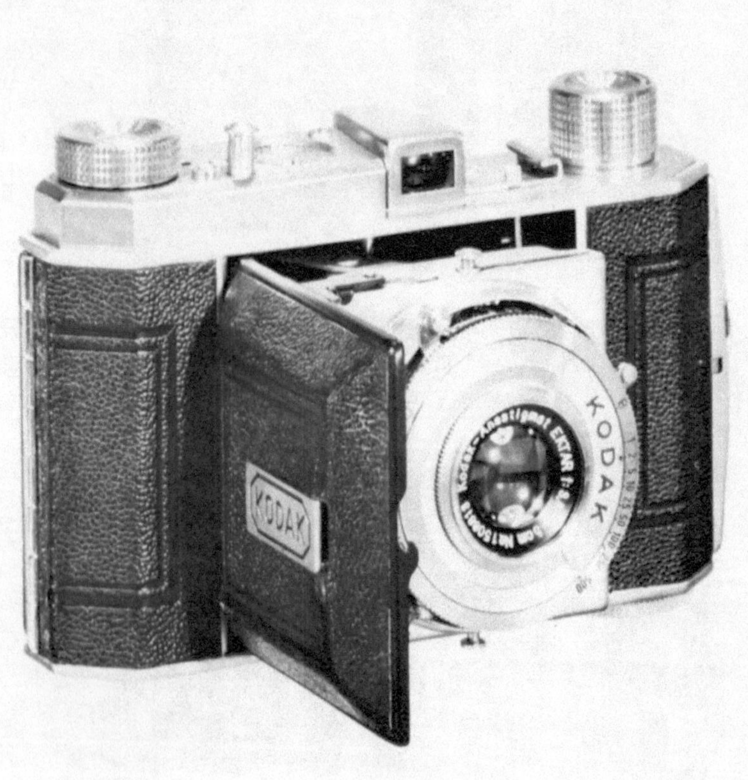

The Retina Ia was a trim evolution of the basic Retina design. The shutter release was incorporated in the body, and an accessory clip was added. Except that it had more chrome trim, it remained faithful to the original. The design ended with the introduction of the C series.

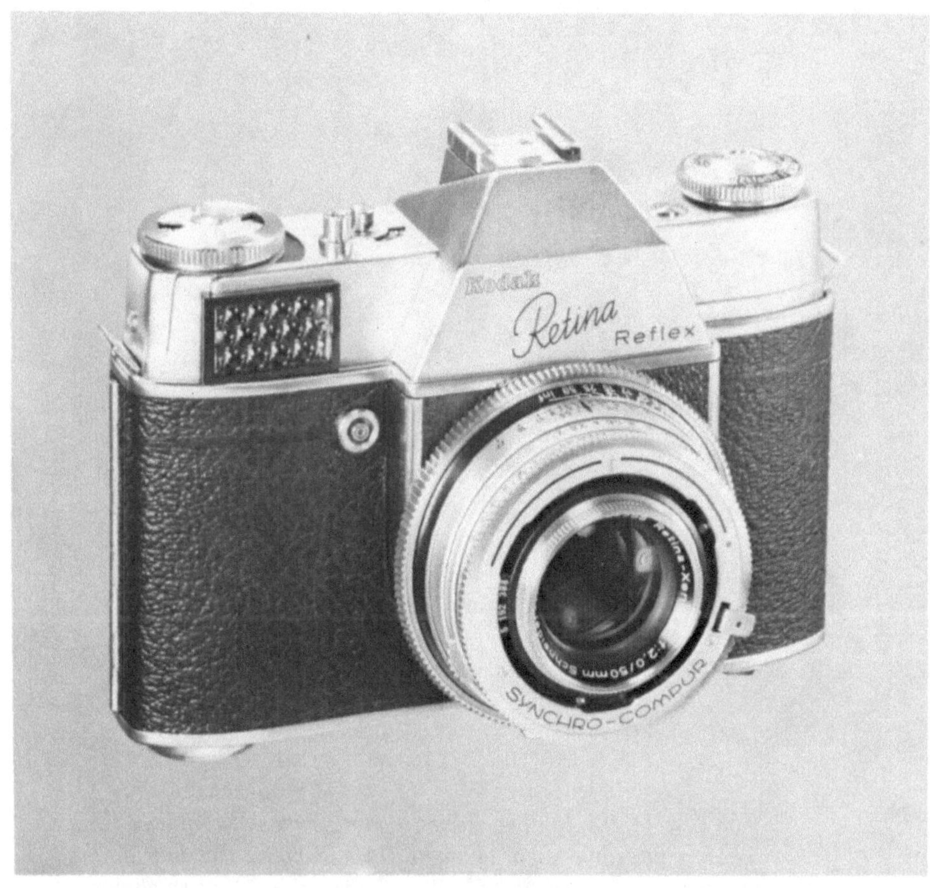

The first Kodak Retina Reflex followed the basic design of the folding Retina series. This first model was a no-nonsense camera, as were most other German-made single-lens reflexes of the time. Unfortunately for Kodak, the Japanese were more innovative in design and quickly took over the market.

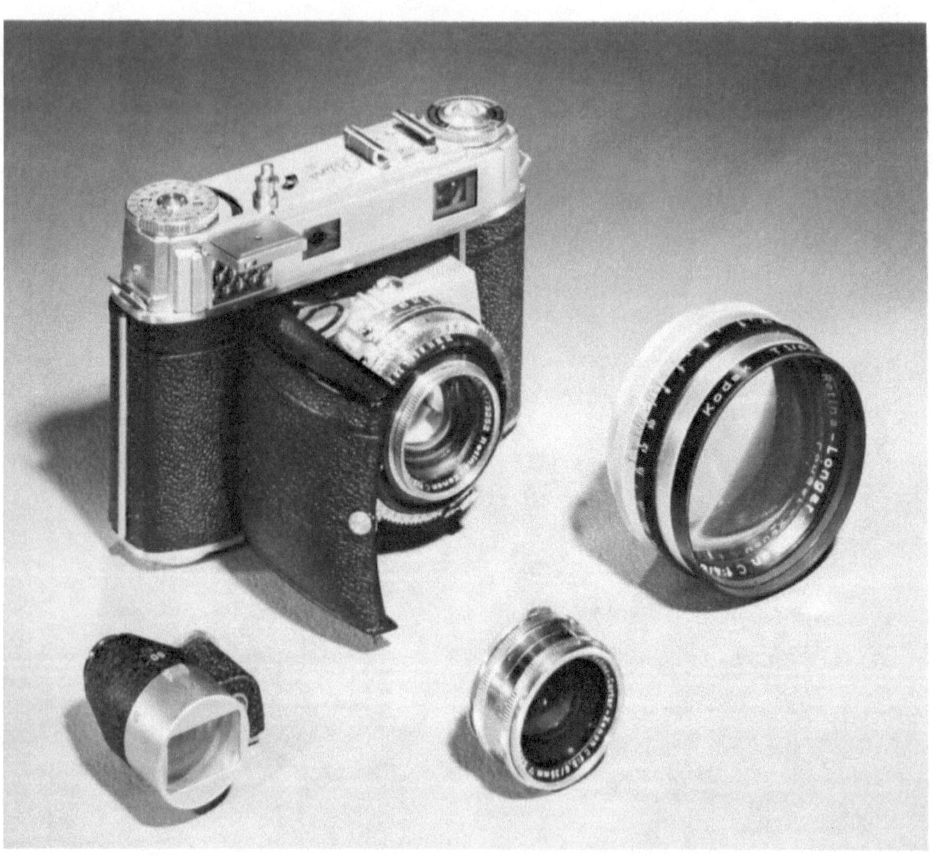

The last of the folding Retina cameras was the model IIIC. There were some slight changes in the camera: the door on the meter was removed and a built-in baffle added. The small c *was changed to a capital* C *in the model number. The lens elements were now interchangeable, and the shutter was fully flash-synchronized, with speeds and diaphragm interconnected for use of EVS numbers. Many people feel that this camera is one of the most difficult of all Retinas to learn to use.*

worthy of its attention and that the economic factor of 35mm sales did not appeal to a large manufacturer primarily interested in expanding film sales. However, while some Kodak officials were no doubt prejudiced against the smaller film size by virtue of background and personal preference, it was never an official or even an unofficial position taken by the company, which went to great lengths to dispel the rumors. But Kodak had its own ace up its sleeve, and in June, 1936, the Kodak Bantam camera series was placed on the market.

Charles Z. Case, one-time manager of Kodak's English associate company, Kodak Ltd., had carefully watched the Leica and other 35mm's develop from playthings into serious instruments, but he was convinced that 35mm had several serious disadvantages which would prevent long-term popularity. Among others, Case felt that the 24 × 36mm picture area was too small, that 20 or 36 exposures in a cartridge were too many, and that, when the camera was set aside, there was no easy, foolproof way to remember exactly which emulsion was in the camera when it was brought out for a picture-taking session weeks later.

Working with the basic 35mm film stock, Case devised a new format which he felt held great promise, and Kodak agreed. Designated 828 (and popularly referred to as 828 Bantam), the new 28 × 40mm picture area was 25 per cent larger than 35mm. The 828 negatives could be used in 35mm enlargers (with a slightly larger mask opening), and color slides remained 2 × 2 inches in exterior dimension for compatibility with 35mm slide projectors. The new film delivered eight exposures a roll and could be knob-transported through a camera to an automatic stop by means of a patented edge perforation (through both film and paper backing) not unlike today's Kodak Instamatic film—one perforation a frame. Edge-numbered, the film used a paper backing for protection against fogging and for easy identification through the camera's small green viewing window in the back.

The first Kodak Bantam appeared in two models: the *f*/6.3 Anastigmat ($9.75) and the *f*/11 Doublet ($5.75)—identical except for lens and price. The body and the detachable back were made of Bakelite, with a decorative corrugated striping effect. Weighing in at 7 ounces unloaded, the Kodak Bantam had a folding bellows, a fixed-focus unit with a simple shutter, giving either snapshot or time exposures. The spool chamber was fitted with a pressure spring which held the small roll of film taut. Once the paper leader was threaded across the camera and attached to the take-up spool, the back was snapped into place and the film wound until it stopped automatically at the first exposure.

Film advance was accomplished by momentarily depressing a release pin which disengaged the tiny film-stop lever from the film perforation. With this system in operation, the green window was needed only for identifying the film (printed on the paper backing) and counting the exposures. A folding metal leg held the Kodak Bantam vertically on a flat surface. An optional optical viewfinder was available for the Doublet lens model (standard on the *f*/6.3 model). Kodak 828 Panatomic film was priced at 20 cents for an eight-exposure roll, which made each picture a fraction of a cent more expensive than 35mm. Kodak Bantams appeared in 1938 equipped with an *f*/8, an *f*/5.6, or an *f*/4.5 lens. The prospective owner could now have his choice of five standard models, as well as one superdeluxe version,

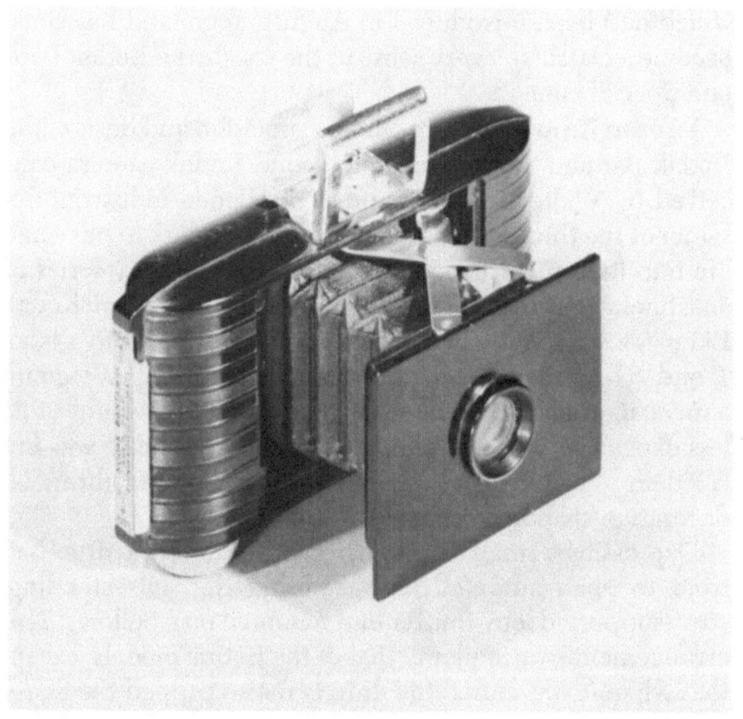

High hopes, never fulfilled, were placed on the Kodak Bantam cameras. Kodak was convinced that the long film loads and the inconvenience of loading would prevent the 35mm camera from becoming popular. The Bantam was designed with a semiautomatic film advance, a larger format, and an eight-exposure roll film with paper backing. The film was 35mm, but with only one perforation a frame to activate the film stop. Although the Bantam was available in a wide variety of models, including the classic Special, the public never took to the 828 film size. Difficulties in film flatness were also encountered with the tiny roll of film, which led to significant research in that area and indirectly to the Kodak Medalist camera.

which had been introduced in August, 1936, and has since become a classic in every sense of the word: the Kodak Bantam Special camera.

An exercise in miniature-camera precision and design, the Kodak Bantam Special was the second Kodak camera conceived by Walter Dowin Teague, the famed industrial designer of the thirties who was retained by Kodak to put some life into its camera line.[5] The first of a brand-new series of anastigmat lenses to come from Kodak, the 45mm Kodak Ektar *f*/2.0 was set in a Compur Rapid shutter (1 to 1/500, *T* and *B*) and headlined the camera's features.[6] Weighing a mere 16 ounces, or little more than twice the weight of its less-expensive brothers, the Special's die-cast body was finished in a baked black enamel, with a cast bright-metal decorative striping to break up the mass.

Depressing a small metal spring cap allowed the drop-bed front to open automatically, exposing the self-extending, strut-supported lens-shutter unit mounted on a bellows. This arrangement was similar to that of the Retina models, except that when it was closed the drop bed also capped the finder housing for protection from damage and dust. The top housing incorporated the eye-level viewfinder and the split-image rangefinder window, the eyepiece of which was adjustable to individual eye variations. The rangefinder was

[5] Teague liked bright work, contours, and shadow effects. His first design at Kodak was the Kodak 127 Baby Brownie camera, and his influence was to mark the end of Kodak's black-crepe era of camera design. Teague's assignment was concerned only with styling; it was left to Chester Crumrine and others to design the optical and mechanical functions to fit the style. Crumrine was responsible for the Kodak Bantam Special, the Kodak Medalist camera, and the K-24 aerial camera, among others.

[6] Ektar is a trade name, not a lens formula, as is often thought. Its design has varied over the years, but always has denoted the best of the Kodak lenses.

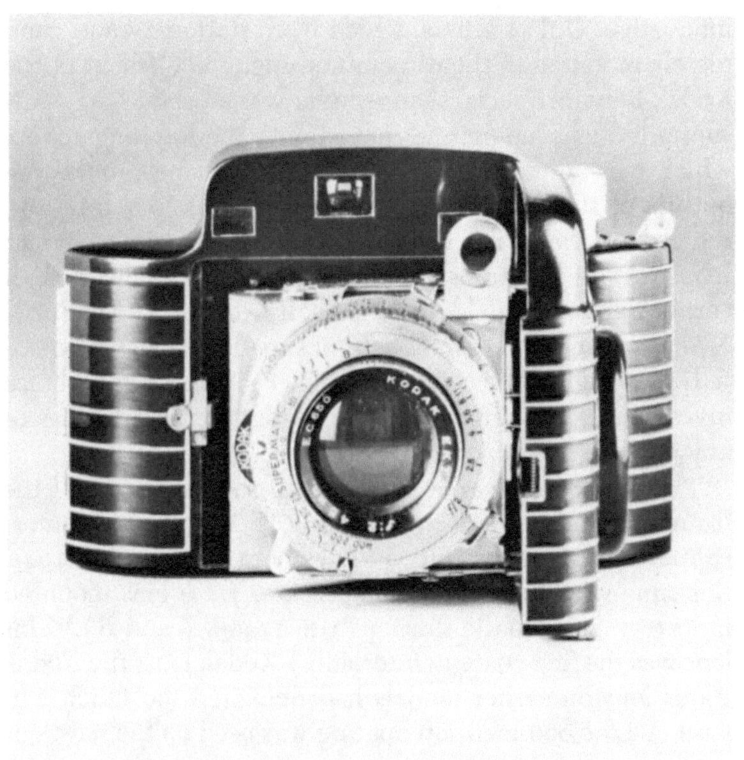

The Kodak Bantam Special is one of the truly great classic American cameras. Its high price ($110) kept it from being as popular in its own time as it was when it went out of production, but its compact size, sleek design, and large format was combined with a needle-sharp lens (later used on the Kardon) to make it one of Kodak's best picture takers. Only minor differences in the winding knob, the lens support, and the shutter identify the various production runs. Early models had Compur and Compur Rapid shutters, but the war caused a switch to the American-made Supermatic shutter (above).

innovative. Unlike others, which were stationary and completely mounted in the camera housing, one element of the Kodak Bantam Special's rangefinder was affixed to the strut-supported lens mount and moved with the focusing action, which was actuated by a radial lever moving around the outside of the shutter mount. Focusing was to 3 feet, and the camera could not be closed until the focus was set at infinity. The film-transport system was essentially that of earlier Bantam models, and the availability of Kodak Super X and Kodachrome film in 828 rolls made its market potential very bright. The camera sold for $110 and was a choice investment, retaining one of the highest trade-in values of any camera manufactured.[7]

The advent of war in Europe in 1939 soon made the Compur Rapid shutter unobtainable,[8] but after a brief interruption Kodak reintroduced the camera in the fall of 1941, this time with the 45mm Kodak Ektar *f*/2.0 lens mounted in its own Supermatic shutter (1 to 1/400, *T* and *B*).[9] This lens was the first coated (internally) Kodak lens; the Kodak Ektar lens on earlier models had delivered too much lens flare. A 1/50,000-inch soft coating was used on the new lens

[7] A few of the Kodak Bantam Special cameras were manufactured after the war. It was officially discontinued in 1947, but prewar models sold after the war for more than their original cost. In the early fifties one of the authors paid $80 for a mint model, and even today, thirty-five years after it was introduced, the Kodak Bantam Special brings $40 to $50 in photo shops, and if the clerk sizes up the buyer as a connoisseur of fine craftsmanship, the price goes up. One Bantam Special recently brought $100 at a photo auction in Los Angeles.

[8] An English torpedo sank the ship carrying the final shipment from Germany.

[9] The Kodak Supermatic shutter was very much like the Compur (both were gear-retard shutters), except that it used a step rather than a continuous cam. On the Compur by setting the speed dial between the marked speeds, one could obtain an in-between speed. Not so on the Supermatic, whose designers felt the step-cam system delivered more accurate speeds.

to produce better contrast and color purity. This model also carried a red index mark on the engraved distance scale for manual-focusing compensation when infrared film was used.

The 828 format reappeared after the war in both the *f*/6.3 and the *f*/4.5 models (until 1947 and 1948, respectively). In 1947 the Kodak Flash Bantam camera was introduced at $49.50; a year later it was the sole survivor of the line. Basically the *f*/4.5 model with flash synchronization added, the Flash Bantam was a simple design with neat, clean lines, and few small-format cameras have ever approached its compactness and ease of handling. In 1953 it was replaced by the Kodak Bantam Rangefinder model (*f*/3.9 Ektanon

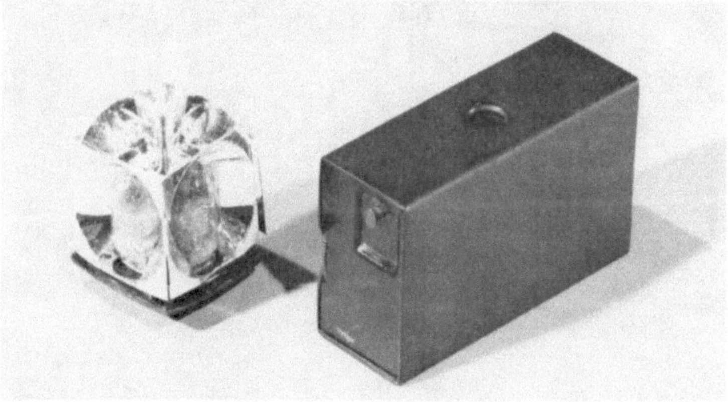

During World War II, Kodak stored its dies for civilian cameras and went into military production full time. One of its products was this little-known matchbox camera of 16mm size (shown above with a flashcube to illustrate its size). The camera was packed with a complete "spy outfit," with developing powders, film, and accessories. Exactly how many of these cameras were made and what the government did with them is unknown.

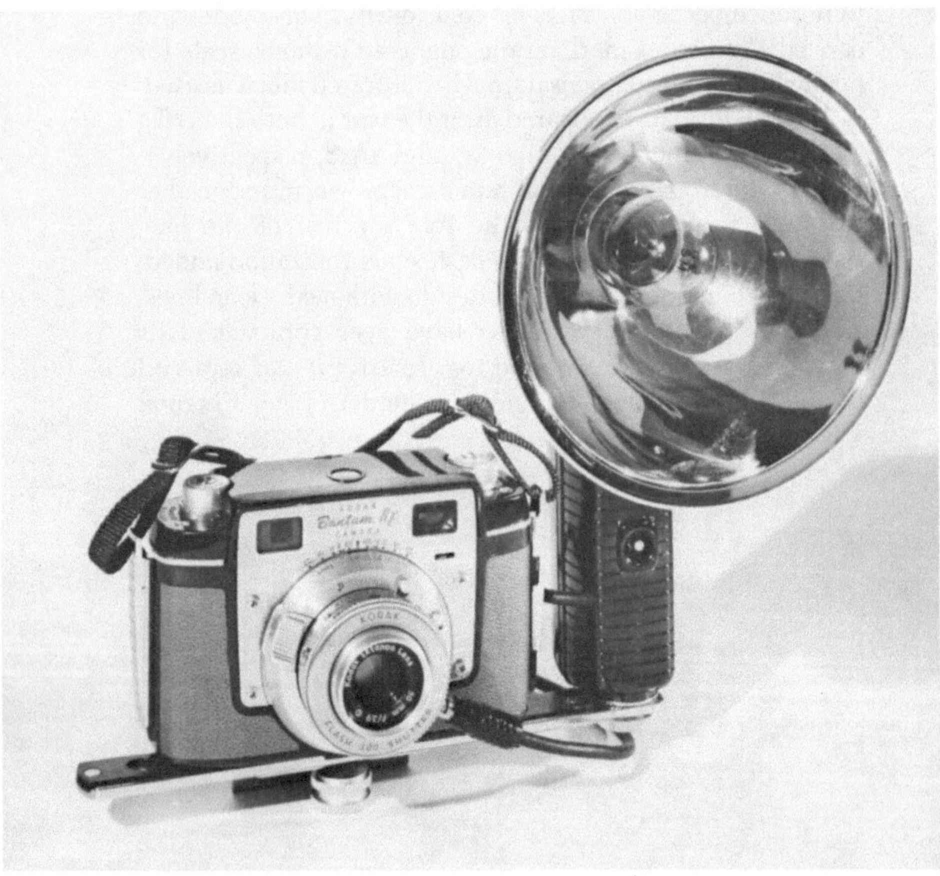

The Bantam RF was the last of the 828 cameras. In quality it could not measure up to the Kodak Bantam Special, but it was the only other 828 camera with a coupled rangefinder. The RF was a neat handful, but it proved to Kodak once and for all that 828 was a film size of the past. Originally listing at $60.25 each, the cameras were finally sold in 1957 by Montgomery Ward at $29.97, complete with flashgun.

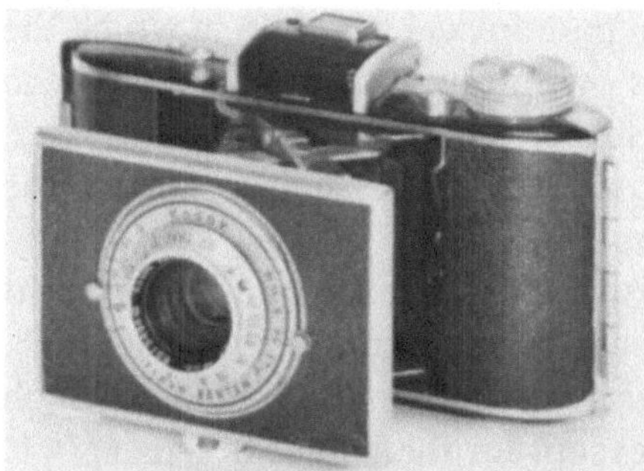

The Kodak Flash Bantam was probably the most popular of all Bantam cameras. It was the only prewar Bantam to be returned to production after the war. Kodak added flash and a coated lens to make the neat, compact model even more desirable. Its small size, combined with an excellent lens and shutter and a semiautomatic film advance, made it a popular seller. Owners of the camera were and still are a fiercely loyal group who regard 35mm with the same disdain that Leica owners reserve for Argus lovers; the inevitable extinction of 828 film will be a cruel blow to them. The 828 Bantams will soon slip into history with the UniveX OO cameras and others that used nonstandard film sizes.

lens and Flash 300 shutter, 1/25 to 1/300, and *B*), a complete design alteration which spelled the end of the 828 line in 1956.[10]

Over the years the 828 film size failed to achieve the popularity which Charles Case had anticipated. Kodak was never successful in attaining wide distribution of the film, though the Bantam cameras were very popular with their owners (especially the Special and the *f*/4.5 Flash models), becoming one of the most revered camera lines of that or

[10] 828 film was a postwar favorite for the inexpensive premium-type cameras, and, while other film manufacturers packaged some of their emulsions briefly in that size, no other serious designs ever appeared from major camera manufacturers until the Fotron in the early sixties. Marketed door to door at around $400 (including five years of "free film"), this product of the Triad Corporation, of Glendale, California, features a built-in electronic flash unit, and an *f*/6.7 lens and uses a sealed drop-in cartridge loaded with 828 film, delivering 10 exposures per roll.

any other period. It was a paradoxical situation in which the film system held back its own growth. The mistake was in introducing the cameras first and then waiting for film sales, an error that Kodak would not make in introducing the Kodak Instamatic system.

The success of the inexpensive Argus 35mm line spurred Kodak to introduce a lower-priced 35mm camera in 1938, the Kodak 35.[11] The 35 was available in three models which were basically the same except for the choice of lens (*f*/5.6, *f*/4.5, and *f*/3.5 Kodak Anastigmats) and price ($14.50, $24.50, and $33.50). Each camera had a noninterchangeable lens and a rim-set shutter (Kodex 1/25 to 1/100, *T* and *B*; Diamatic 1/25 to 1/150, *T* and *B*; or Kodamatic 1/10 to 1/200, *T* and *B*) in a rigid mount. Bulky but sturdy in design, all the Kodak 35's were equipped with folding optical viewfinders similar to the one used on the Kodak Bantam *f*/4.5. The major operational distinction in the three models was in shutter interlock: the *f*/5.6 model had none, but in the other two cameras winding the film cocked the shutter. The shutters were not synchronized for flash, but all had built-in delayed-action self-timers.

As we look back today, the early history of the Kodak 35 was a great tragedy, for it fared very poorly in competition with the Argus C series. The metal and plastic Kodak 35 was described by one critic as "the design of a blacksmith." While the Argus C series was no better-looking, it sold rings around the Kodak 35's with hardly any effort. Sales and production records were not preserved, and so it is impossible to state now how unequal the sales of the two camera

[11] The lead time necessary for a company the size of Kodak would indicate that the cameras had long been in the planning stage, but the success of the Argus was obviously an impetus to the introduction of the Kodak 35 camera.

lines were, but reliable industry sources have indicated to the authors that the Argus C series outsold its Kodak competition by approximately 100 to 1. Neither Argus nor Kodak ever determined exactly why this great disparity existed. Kodak had the quality and the name, but Argus had the lower price, and that seems to have been the determinant. Industry insiders insist that the absence of flash synchronization from the Kodak line could not have accounted for the great difference in sales.

The Kodak 35 camera was a well-built, functional, and dependable performer with a lens quality which remained constant, unlike the varying Argus Cintar, which had its ups and downs, depending on the manufacturer. Yet the Kodak 35 offered but token competition in the sales market. In 1940, soon after the C-3 was introduced at $35 with coupled rangefinder and flash synchronization (both features lacking in the Kodak 35 line), Kodak brought out a coupled rangefinder version of the Kodak 35 *f*/3.5, but $50 was the lowest price at which it could be marketed.[12]

The Kodak 35 Rangefinder model had the same basic design as the previous models, except for a metal housing across the top which contained the viewfinder and the split-image–rangefinder mechanism. Bracketed on the right side of the lens mount was an 11/16-inch milled wheel which meshed with a milled ring on the front of the Kodak Anastigmat Special lens. Turning this wheel focused the lens from 4 feet to infinity, but its addition, along with the rangefinder housing, which extended to the milled ring of the shutter on the opposite side of the lens, made it difficult to

[12] Even after World War II the Kodak 35 was more expensive, selling at $86.75 without case or flash, while the C-3 with case and flash cost only $66.50.

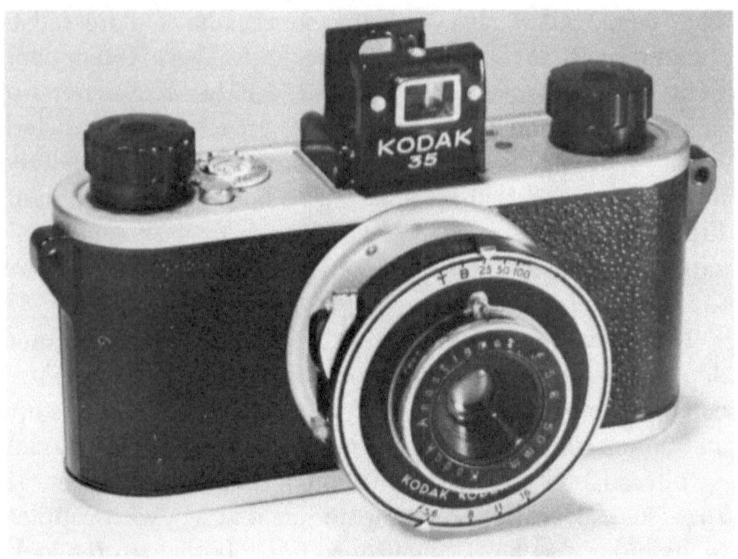

Available in three models, the Kodak 35 line was introduced to compete with the fast-selling Argus A and C series. This particular model featured an f/5.6 lens and proved a sturdy, dependable performer.

adjust the rim-set shutter—there was not sufficient room to get a good grip. The sharp teeth of the focusing wheel could numb a finger after a lengthy and rapid picture-taking session; the arrangement was not designed for rapid focusing. The rangefinder housing on the left side tended to interfere with a comfortable grip. As a result, the camera was best held cradled in the user's left palm steadied between his thumb and forefinger. This model seemed destined for oblivion until the army bought a quantity (said to be in the neighborhood of 50,000) during the early part of the war.

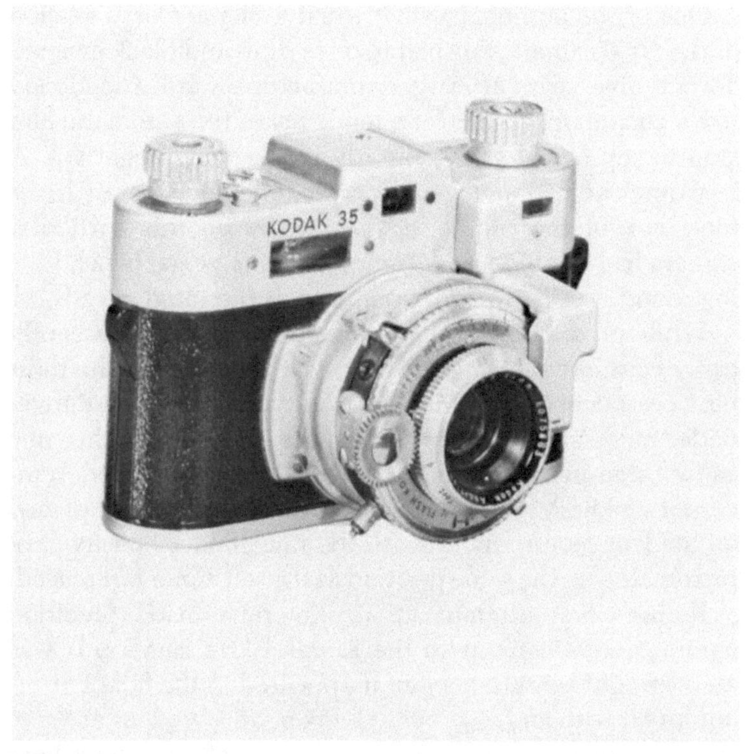

"The camera of afterthoughts" seems an accurate description of the Kodak 35 with rangefinder. The body was basically that of the original Kodak 35's. The rangefinder was added primarily because of the sales pace set by the Argus C-3. The prewar models had self-timers which were replaced by flash synchronization after the war, along with coated lenses. Postwar models also had plastic winding knobs on most production runs. For all their jumbled appearance, the cameras were excellent picture takers and extremely sturdy. While they had a loyal following, it is doubtful that total sales ever threatened those of the remarkable Argus C-3.

One of the authors briefly owned a Signal Corps version of the *f*/5.6 camera, finished in olive drab and black enamel. Twenty-five years after its manufacture it still functioned like a champion, except for a tendency of the automatic film stop to slip occasionally, slightly overlapping negatives. A few quick adjustments with a pipe knife and a screwdriver took care of the problem. The shutter on this particular camera had retained its accuracy over the years; bench testing found it to be within 5 per cent of the rated speeds.

While production of the entire Kodak 35 series was officially continued during the war, manufacture for the most part ceased in 1942. Only the *f*/4.5 and the *f*/3.5 Rangefinder models were revived after the war, and neither one caused stampedes to camera stores. The postwar models incorporated flash synch by means of an ASA flash post placed on the lens mount underneath the rangefinder housing, approximately in the same position as the self-timer it replaced.

Kodak's first attempt at an American-made precision 35mm camera came with the Kodak Ektra camera. It was an overnight sensation when it appeared in the fall of 1941, equipped with an *f*/3.5 lens at $235 or an *f*/1.9 lens at $300. No one knows exactly how many cameras of this model were manufactured before the war brought production to a halt, but informed sources at Kodak place the figure at approximately 2,000.

In concept the Kodak Ektra was to incorporate all the design innovations of previous cameras with the over-all "grand design"—a series of movie and still cameras which would in themselves represent a landmark but would be particularly notable for specific innovations.[13] Joe Mihallyi

[13] The Kodak Cine Special 16mm movie camera that started the series was designed and in production within ninety days. It was a model shop

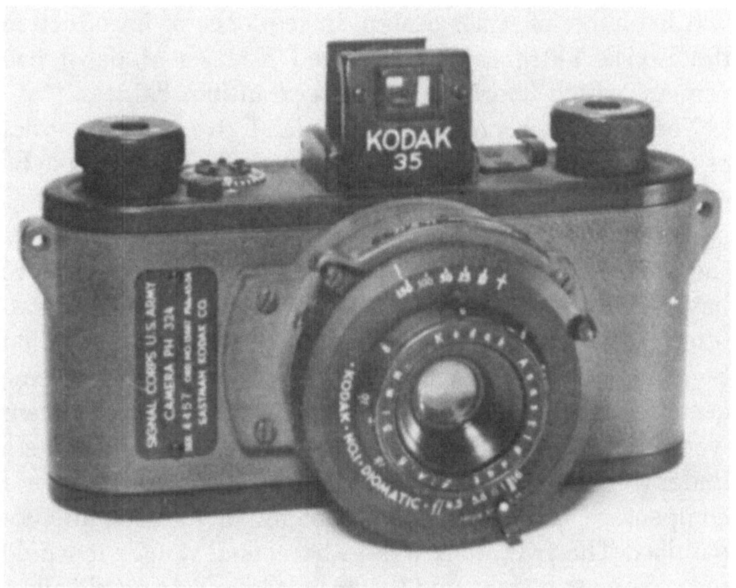

During the war years Kodak produced an f/4.5 version of the Kodak 35 for use by the United States Signal Corps. The olive-drab plastic body was trimmed in black, and while many cameras were available to the public as war surplus in the late forties, this model is now a genuine collector's item.

operation and each one was literally hand-built. It was followed by the Kodak Sound Special projector, an exercise in "overengineering" to prove to Hollywood that 16mm sound-on-film had the potential of good sound. This projector used a viscose drive which gave it a never-excelled smoothness and sound quality. Unfortunately, it was impossibly expensive to manufacture at the time and refused to operate in cold weather. The automatic-exposure Kodak Super 620, the Kodak Bantam Special's rangefinder, and the Kodak Medalist's film flatness were further steps in the series.

was in charge of series design. In sequence of introduction the Kodak Ektra camera followed Kodak's Medalist 620 camera, which had been an exercise in film flatness.[14]

The original proposal for the Kodak Ektra was for a series of cameras to be known as the Kodak Super 35. It was to be capable of using interchangeable film magazines, or "backs," each of which would indicate to a built-in exposure device the kind of film contained within. Each back was also to have its own rapid-advance lever wind and was to be hand-fitted at the factory to the camera body with which it would be used. The split-image rangefinder was to have extreme accuracy and brilliance. The rigid interchangeable lens would not require a separate viewfinder, since the camera finder would zoom internally for use with various lenses and compensate both for vision problems of its user and for parallax. The *f*/1.9 lens was to have dual range: it would focus down to 3½ feet, and then a button release would allow closer focus to about 1½ feet, a distance at which the rangefinder would continue to function. This dual range would permit close-up pictures without the use of close-up lenses, which tend to reduce quality.

[14] Most of the basic research in film flatness was done for this camera because of problems encountered in trying to build a folding camera with an *f*/3.5 lens. The problem rested in the fact that roll film was designed to overcome the curl of the roll by a slight tendency to curl with the edges inward (toward the lens). At large lens openings, the curling made the edges unsharp; the same problem was true with press cameras, which were generally unsharp at large openings because the sheet film was not completely flat in the holder and even less so in film packs. The resulting design of the Kodak Medalist camera gave the film a two-way curl, thus overcoming the problem. The Medalist was one of the sharpest picture takers ever produced by Kodak and introduced the concept of interchangeable backs, at least to the extent of allowing a ground-glass back to be substituted for the regular back, and a cut-film holder to be used in place of roll film. While the idea was not new, most cameras, such as the Kodak Recomar, used the opposite approach. Kodak had never before produced a roll film attachment for a plate camera.

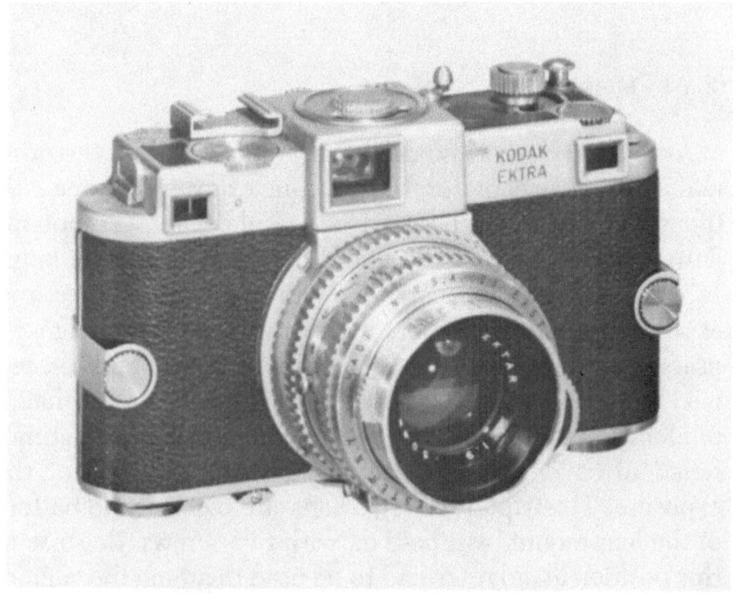

The Kodak Ektra represented the height of American camera design and reached a level of quality which will probably never be equaled. The camera was far ahead of its time, with features that were not found in other cameras for almost 20 years. Unfortunately, it was introduced in 1941, and production ceased before it really got started. The exact number manufactured is unknown, although informed sources estimate fewer than 2,000. For all its advanced features, it lacked a few that might have been expected. For example, the flash was synchronized by an external attachment, although the flashgun had a small light bulb that could be used for setting the camera at night. The Super Kodak 620 had a combination range- and viewfinder, but the finders were separate on the Ektra—although they were infinitely adjustable. The camera's weakest feature was the shutter, which was never perfected. Surprisingly, the tripod socket looked somewhat like an afterthought and was a constant source of annoyance. The model shown above has the internally coated f/1.9 lens. The small crank at the top is the self-timer. Note that the exposure release is on the left. The rapid-advance lever is on the back, also at the left.

As produced, the Kodak Ektra camera followed the original concept except for the built-in exposure device. Although it had everything anyone could want (except for flash synchronization), the Ektra contained its share of bugs. It was a left-handed camera, with rapid-wind lever and exposure release on the left, an awkward position for many users. The delayed-action timer was a small crank placed next to the exposure release. The focal-plane shutter had a tendency not to "cap"—the second shutter blind sometimes failed to cover the first, leaving a slit at the end of the exposure. The tripod socket, a separate block at the bottom of the lens mount, was held on with two screws. Because of this position an adapter had to be used to attach the camera to a tripod, and users found that the screws tended to loosen.

While the interchangeable backs could be removed without losing any exposed pictures, taking them off was a chore. The thumb wheels which held the back to the camera were difficult to use when properly tightened down. The camera's outstanding success was its complicated rangefinder, which was operated by a large wheel on the lower right side of the camera front. The wheel engaged a rather hefty gear on the lens mount for focusing. Brilliant in use—and also expensive to manufacture—the rangefinder was separate from the viewfinder. A favorite demonstration of the rangefinder's accuracy was to hold a pencil about 18 inches from the customer at a 45-degree angle. The customer focused the camera on the center of the pencil—or tried to; the two halves would not focus because the rangefinder could discriminate between the difference in the top and bottom angles of the pencil in that position.

The Kodak Ektra was equipped with a choice of two

excellent, basic lenses. The *f*/3.5 was the identical lens previously used on the Kodak Rangefinder 35mm, a "stretched" (reworked) Tessar formula. The *f*/1.9 lens was only a slight change from the *f*/2.0 used on the Kodak Bantam Special.

The Ektra dies and parts were stored during the war years. In 1946 the question of whether production should be resumed was carefully considered. When Kodak's marketing and sales personnel estimated that it would cost $700 retail to reintroduce the Ektra, at a time when the Leica was selling at around $400, it became obvious that this would be an impractical move (as Bell & Howell found out with its Foton), and the dies were left in storage; Kodak does not dispose of such. The Kodak Ektra immediately assumed the proportions of a legend: it was prewar, precision, and scarce.

Marketing research was but a small part of Eastman Kodak Company's over-all planning before World War II. As with most other companies, it was not so much what the consumer wanted as what Kodak could make that found its way to dealer shelves. Before the war emphasis was placed on camera features, as in the "grand design," rather than on what customers desired. By 1947, however, a separate department had been set up at Kodak to begin developing the tools and techniques of market research, and Kodak camera design entered a new phase. A few models of the Kodak 35 and Bantam series held the line until 1949, when the first Kodak Pony camera appeared in an 828 model at $29.95. The 35mm version followed a year later at $34.75.

Both Kodak Pony cameras featured a 51mm *f*/4.5 Anaston lens and the Kodak Flash 200 shutter, a cocking type with speeds of 1/25 to 1/200 and *B*, set in a collapsible mount. Mainly plastic with brushed-chrome metal parts

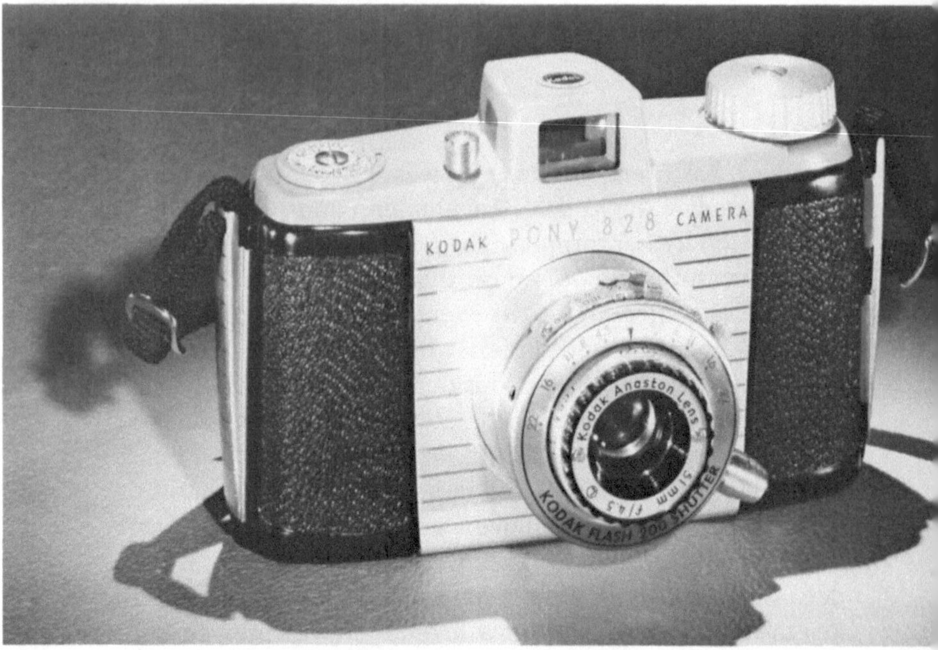

KODAK PONY 828 CAMERA
Kodak Anaston Lens
51mm f/4.5
KODAK FLASH 200 SHUTTER

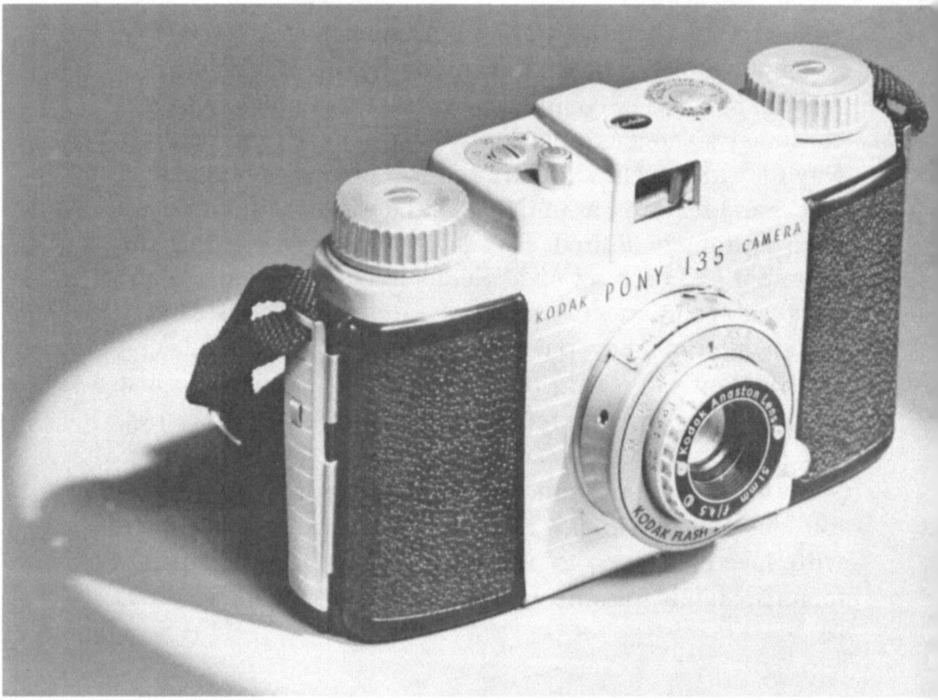

KODAK PONY 135 CAMERA
Kodak Anaston Lens
51mm f/4.5
KODAK FLASH

where required, the cameras were virtually alike in appearance except for the viewfinder housing. The 828 model had a simple optical finder centered on top and placed between the film-wind knob and the film-reminder dial. The 35mm version had a raised housing containing the viewfinder, the exposure counter, and the film-reminder dial, all set between the wind and rewind knobs. Both models had a body shutter release. The first Pony cameras were very popular, and the 828 was produced until 1959, the last of the 828 cameras manufactured by Kodak. The Pony 135 was replaced in 1954 by the Model B, which was in turn superseded in 1955 by the Model C. While the B featured a 44mm *f*/3.5 lens in a collapsible mount, the C differed mainly in its use of a rigid lens mount, with a top shutter speed of 1/300. The C lasted until 1958.

The year before the C was discontinued, the Kodak Pony II (*f*/3.9 Anastar lens, EVS settings of 9.5 to 15) and the Pony IV (*f*/3.5 Anastar lens with Flash 250 shutter, 1/30 to 1/250 and *B*, EVS settings 8.5 to 17) made their appearance. With the exception of lens-mount design and provision for plug-on flash units, the II and IV remained true to the Pony styling.

With the retirement of the Kodak 35 Rangefinder camera in 1951, Kodak brought forth the Signet camera series, which began with the Kodak Signet 35, priced at $95 (soon to drop to $75).

The Pony line was Kodak's first postwar attempt at a low-priced series of 35mm and 828 cameras. Although small and inexpensive, the 828 (top) again proved the unpopularity of the format, while the Pony 135 (below) was a solid hit from its introduction. Except for film format, these two original models were identical, each containing an f/4.5 Anaston lens in a Flash 200 shutter.

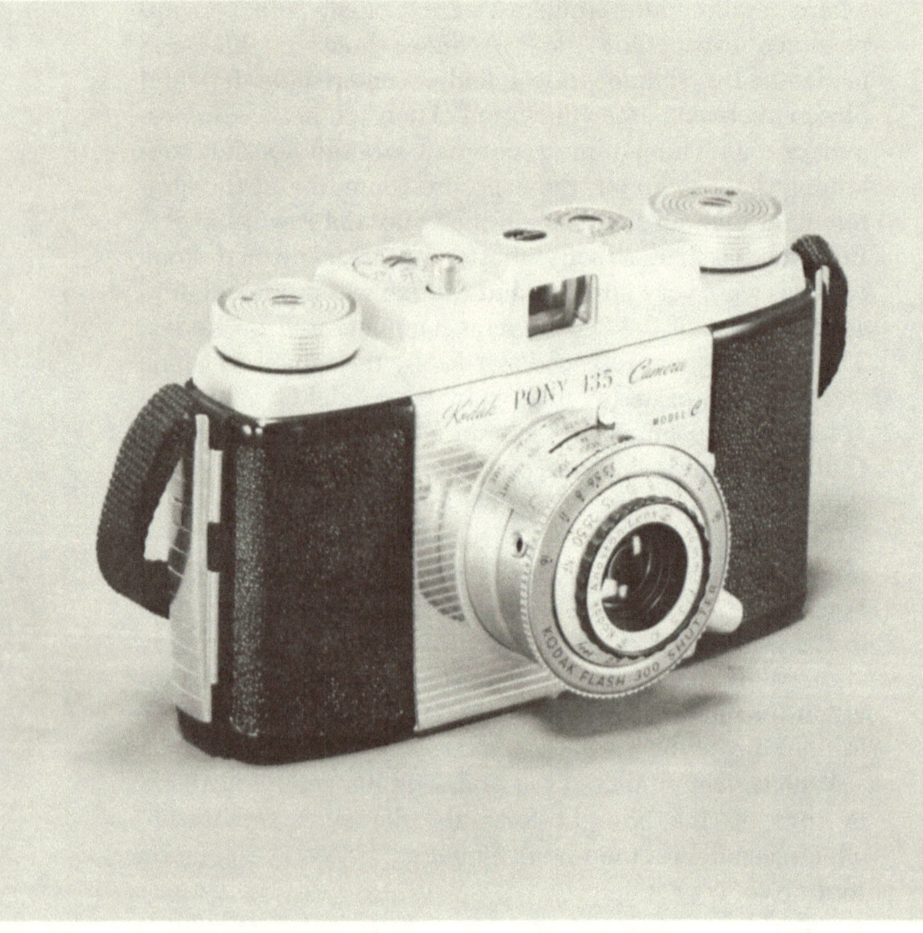

At least one design feature of the Kodak Pony 135 Model C corrected a problem that had plagued users of earlier Pony cameras. The previously collapsing lens mount was made non-collapsible with the simple addition of a metal plate. An f/3.5 lens was added, and the shutter was extended to 1/300 second.

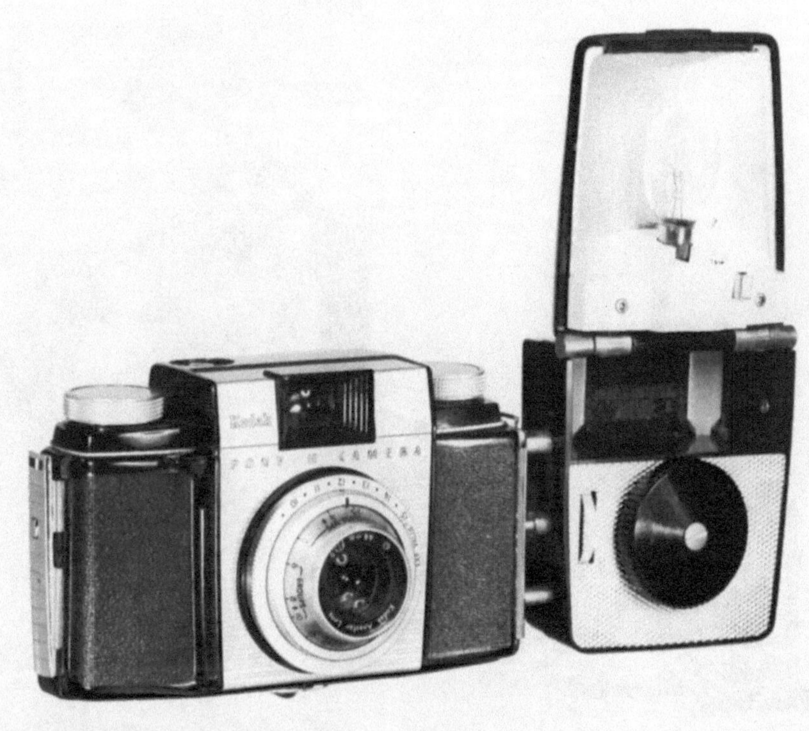

This Pony II is shown with the Generator Flash, which was designed to eliminate the problem of dead batteries. The camera has no marked shutter speeds or f-numbers. The exposure is set by a single EVS number taken from a card inserted on the back of the camera. A different card was used for each type of film and gave exposures much as the charts found in film instructions. The system was simple but frustrating to serious amateurs.

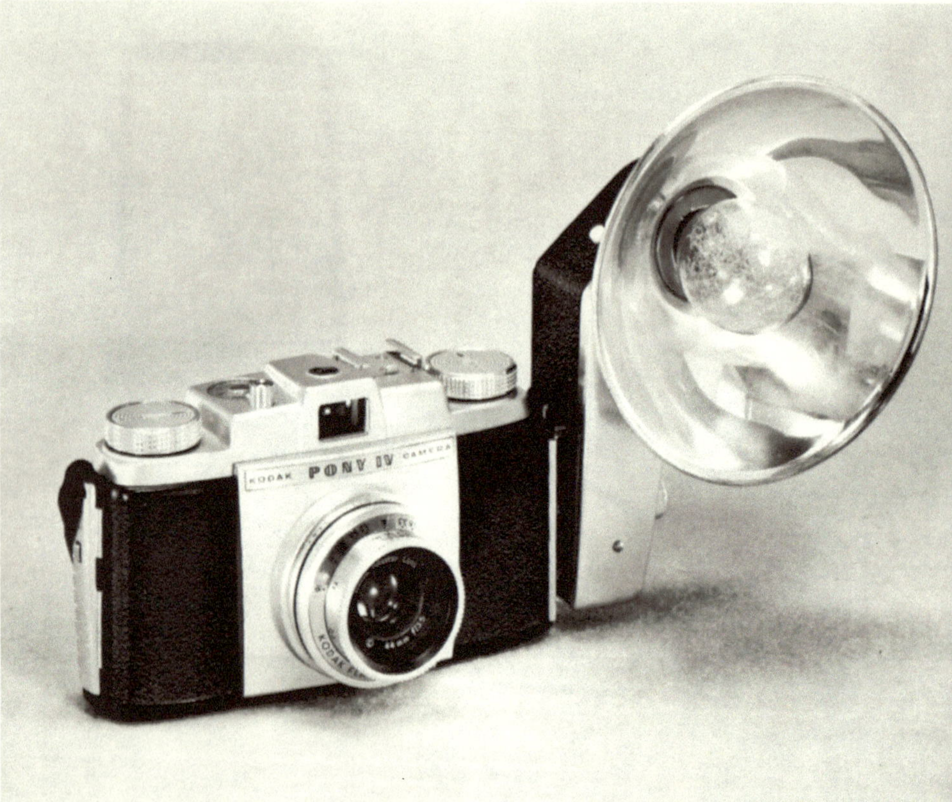

The Kodak Pony IV sported an f/3.5 lens and five shutter speeds which were set by means of an EVS number taken from a chart on the back of the camera. In this series of Pony cameras Kodak attempted to popularize the use of the simple exposure-value system. But camera buyers seemed to belong to one of two groups—those who did not want to set anything and those who wanted to set everything. EVS passed from the scene, as had earlier efforts at simplifying exposure calculation.

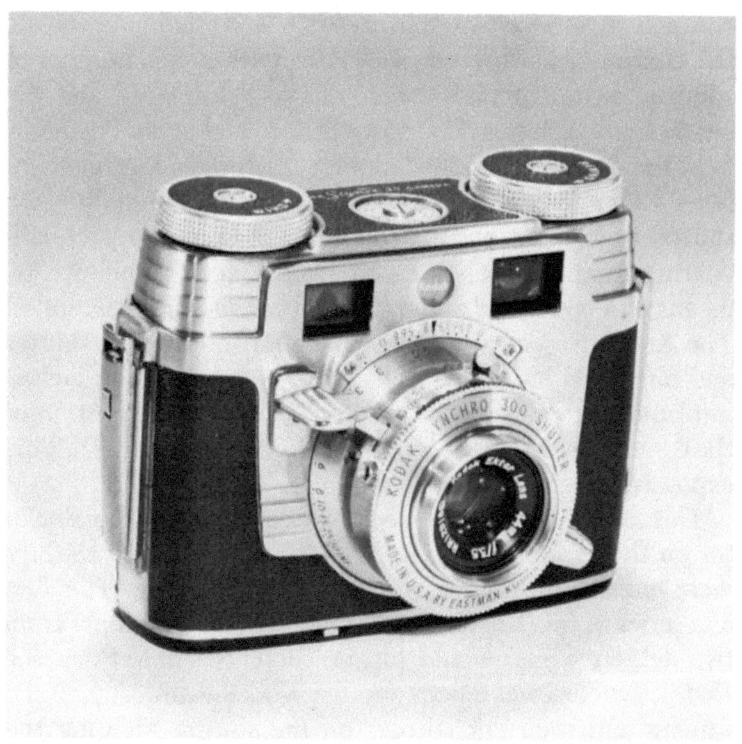

The Kodak Signet 35 was a clean, original design. Small and lightweight, it was exceptionally sturdy and was said by Fortune *magazine to have the best lens of any 35mm camera on the market at the time. The lens was mounted on a ball-bearing ring advertised to contain 50 balls, although it actually had 51. In spite of its sharp-cutting Ektar lens, the camera had one picture-taking fault—the Synchro 300 shutter. To market the camera at a reasonable price, costs had to be shaved at some point, and the decision was made to economize on the shutter. While it was easily repaired, usually with lubrication, it was less than totally reliable.*

Made of die-cast aluminum with a Kodadur covering, this compact handful took its styling cues from a number of sources, including both the Kodak 35 it replaced and the newer Pony cameras. The first of the Kodak 35mm cameras to utilize a 44mm lens, the Signet 35 featured a four-element *f*/3.5 Ektar lens in a set-and-release Kodak Synchro 300 shutter (1/25 to 1/300 and *B*), all housed in a rugged ball-bearing mount which focused from 2 feet to infinity by means of a combined superimposed range- and viewfinder. The Kodak Signet 35 did not have an interlocked shutter and film wind, but once the shutter was tripped, it locked until the film was wound. Moving a lever on the bottom of the camera overrode the lock, allowing an intentional double exposure to be made once the shutter was recocked.

This model also introduced a slide-rule exposure calculator on the camera back. As in the Pony series, all settings were made from the top of the rigid lens mount. This first of a series of five Kodak Signet models suffered mainly from two defects which would plague the entire line. First, the Kodak Synchro was a poor shutter; Kodak overdesigned the camera and then cut corners on the shutter to bring the cost down to a competitive level. Second, the camera was more attractive on the shelf than in the hand.

The metal-and-plastic Kodak Signet 40 camera appeared in 1956, priced at $74 ($65 a year later, including flashgun). It had a noninterchangeable 46mm three-element *f*/3.5 Ektanon lens set in a Kodak Synchro 400 shutter (1/5 to 1/400 and *B*). Utilizing the smooth ball-bearing focusing mount common to all Kodak Signets, it focused to 2 feet through a combined range- and viewfinder operated by moving a knob around the lens barrel. It was the first of the line to feature a quick-loading system; the user simply dropped the

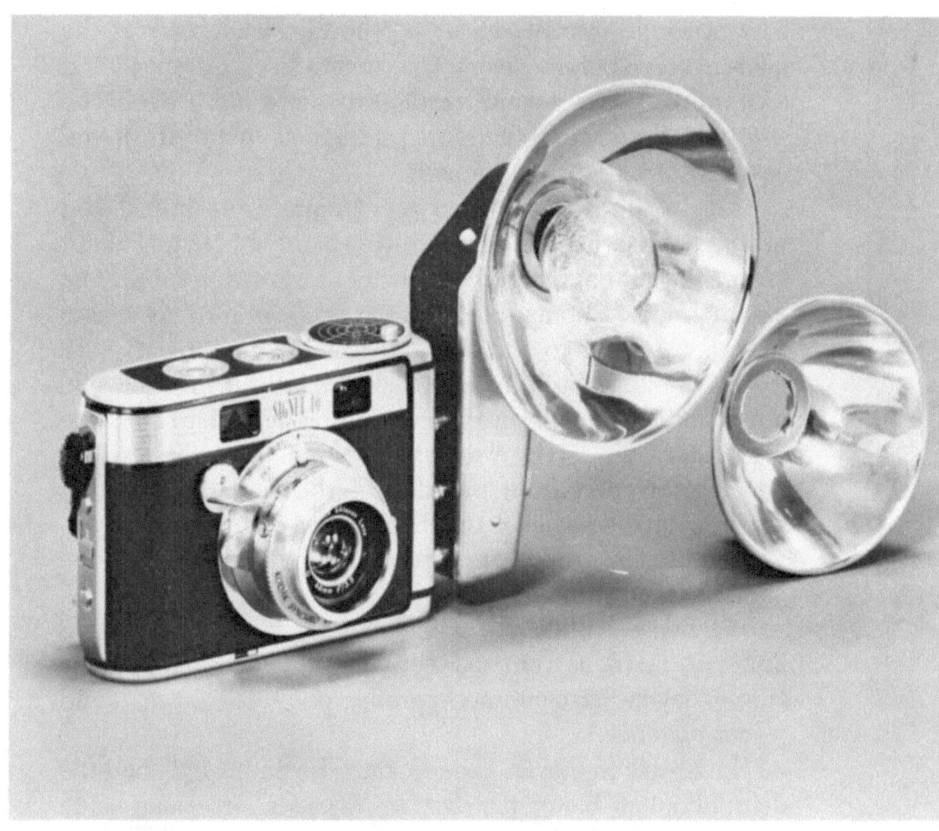

The Kodak Signet 40 was one of the first domestic 35mm cameras with a built-in rapid-wind film advance (Universal had provided one as an accessory for its Mercury in the late thirties). The Signet 40 had a rangefinder and an Ektanon f/3.5 lens. In some respects it was a poor man's Signet 35 with a less-expensive lens but modern refinements. It lacked the classic appeal of the Signet 35, and while the latter is fondly recalled, almost no one remembers the 40. The smaller reflector at the right was used with M2 bulbs.

film cartridge into place, laying the film leader across the take-up spool before closing the camera back. A rapid-wind lever and a rapid-rewind handle expedited film travel. The shutter suffered from the same problem as that of its predecessor: it was poorly designed.

The Kodak Signet 30 and 50 cameras were introduced in 1957. The latter was an interesting camera with just about everything a beginner could want designed into it. The 44mm Kodak Ektanar *f*/2.8 lens was set in a Synchro 250 shutter (¼ to 1/250 and *B*), and had EVS settings from 4 to 17. It had a built-in light meter on the top right side which indicated exposure for films rated ASA 10 to 400. The film index was set by revolving a small horizontal wheel which protruded from the side of the camera. A single-stroke rapid-wind lever on the bottom of the camera advanced the film, and the exposure counter was placed on the camera face at the bottom right of the lens, beside the rewind release. While the Kodak Signet 50 had no rangefinder, it did focus from 2½ feet to infinity, and the luminous-frame viewfinder contained parallax pointers for the 3- and 5-foot distances.

The Kodak Signet 80 camera, introduced in 1958, had the distinction of being the last of Kodak's American-made 35mm cameras to feature interchangeable lenses. Priced at $129.50, it met with resounding rejection by the camera-buying public. It was equipped with a 50mm *f*/2.8 Ektanar lens in a bayonet mount, with shutter speeds ranging from ¼ to 1/250. The 80 also had a bright-frame range- and viewfinder, a built-in EVS meter, and a two-stroke rapid film advance. Two additional lenses were provided for the 80, a 35mm wide-angle and a 90mm telephoto, but each used a rather cumbersome auxiliary viewfinder mounted on the

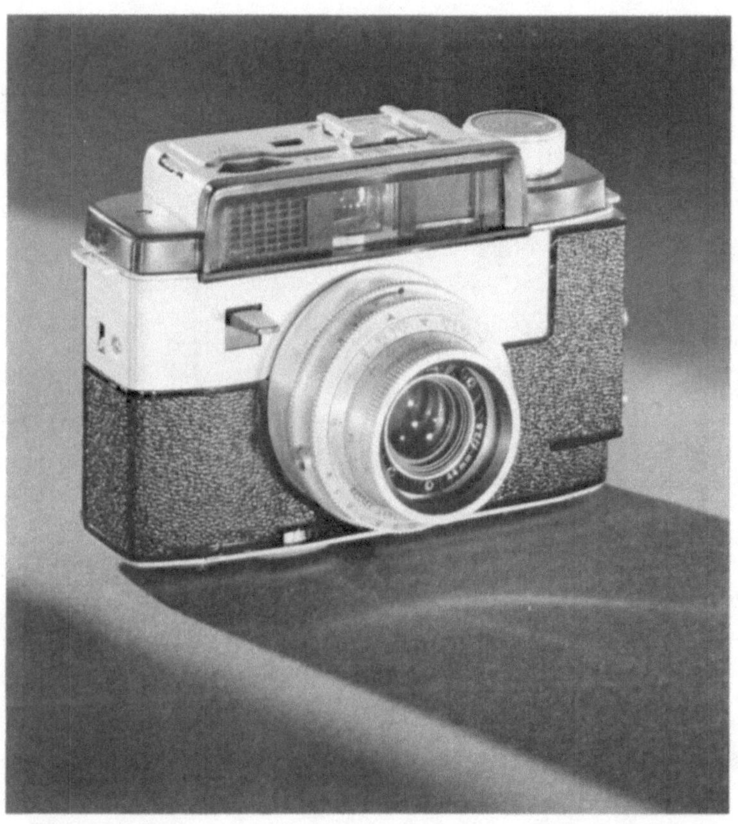

The Kodak Signet 30 and the 50 shown above were identical cameras and one of Kodak's more handsome designs. The 50 differed from the 30 in that it had a built-in photoelectric exposure meter—one of the first Kodak cameras to incorporate such a meter. Both cameras had f/2.8 Ektanar lenses, bright-frame finders, and eight-speed shutters with a top speed of 1/250. They also had an unusual nonthreading loading system and single-stroke film-advance levers. The flashgun was screwed into the side of the camera.

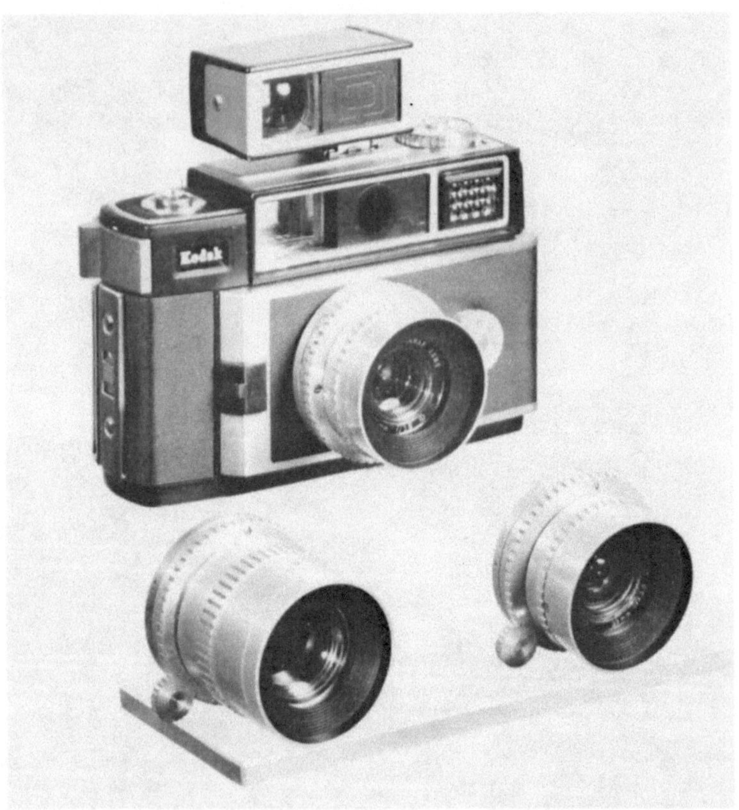

The last American-made interchangeable-lens 35mm was the Kodak Signet 80, a modern design of pleasing lines. While it featured a photoelectric meter, rapid wind, easy loading, and a bright-frame finder, it was comparatively high priced, particularly when compared with imports with equal features. Sales were disappointing and a signal to Kodak that the days of the sophisticated American 35mm were over. In 1959 the camera sold for $129.50 list; two years later Montgomery Ward sold the remaining stock for $59.50.

camera top. A great disappointment over the counter, it was discontinued in 1962.

The final American 35mm design to carry the proud Kodak trademark, the Kodak Automatic 35, appeared in 1959. In 1960 a motorized version appeared, the Kodak Motormatic 35. It had a Synchro 80 shutter[15] and a 44mm *f*/2.8 Kodak Ektanar lens combined with a built-in meter for total automation. Initial pressure on the shutter release caused it to strike against the meter needle, setting the lens opening. Continued pressure tripped the shutter at this preset opening. Since there were no needles to match or line up, the user needed only to point the camera and compose and take his picture. The viewfinder contained a luminous or projected frame, as well as signal indicators for flash setting and low light level and a zone-focusing reminder. The rapid film advance, exposure counter, and rewind release were identical in design and placement to those of the Kodak Signet 50.

The Automatic 35B, introduced in 1961, featured an automatic flash shutter. Setting the flash-guide number on the lens mount and focusing between 5 and 25 feet automatically adjusted the lens opening. The 35F, which appeared in 1962, accepted AG-1 flashbulbs in its built-in flash unit, and the 35R4, introduced in 1966, provided for automatic rotation of flashcubes. The Kodak Motormatic was also available in the F and R4 models with a film transport providing 10 shots per winding. Prices varied somewhat on the different models, but all were within the $90 to $120 range. Production on both the Automatic and the Motormatic models was halted some time before their offi-

[15] The Kodak Automatic had a two-speed shutter (1/40 and 1/80); the Motormatic had a four-speed shutter (1/40 to 1/250).

cial discontinuance in 1969. Even with its quality name behind the product, Kodak could not compete feature for feature on a dollar basis with the foreign imports.

Kodak's most innovative design period had ended with the Kodak Ektra camera, and while its postwar styling was more consumer-oriented, sales were hardly spectacular, except for the original Pony 828 and 135 models, and they were beginners' cameras at best. Although the company produced a healthy line of accessories over the years, much the same could be said about those products, such as the slide projectors. The first Kodaslide projector, introduced in January, 1937, at $48.50, used 200 watts of illumination and

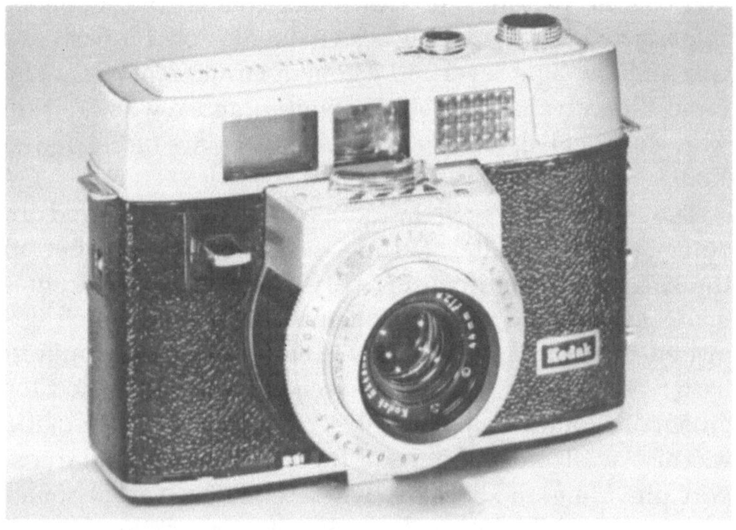

The first of the last American 35mm cameras, the Kodak Automatic 35 had an f/2.8 lens, zone focus, and simple exposure control. It proved to be an immediate sales success even in the face of increasing foreign competition.

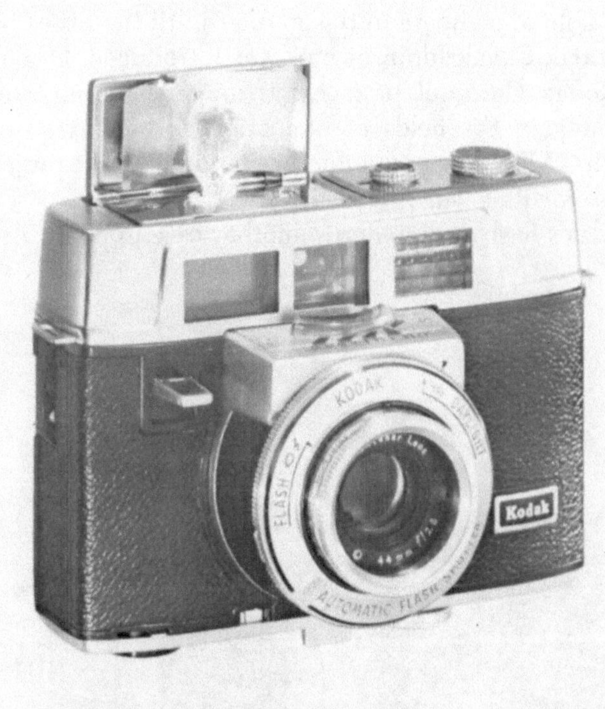

The Kodak Automatic 35F was one of the first 35mm cameras to have built-in flash completely housed in the camera body. This feature was also available in the Motormatic model. Though it did not have a rangefinder, it suited the picture-taking needs of most beginning photographers. Like its predecessor it had a 44mm f/2.8 lens, and it was fully automatic.

natural-draft ventilation. It sold fairly well, but Kodak was never able to compete in this market until the fifties, when the Kodak Cavalcade projector was introduced. Of course, the Kodak Carousel projector is an outstanding success, especially in the fields of education and industry (even though GAF-Sawyer manufactures and sells more projectors because of its lower-priced line).

Kodak's flash equipment is another case in point. Attrac-

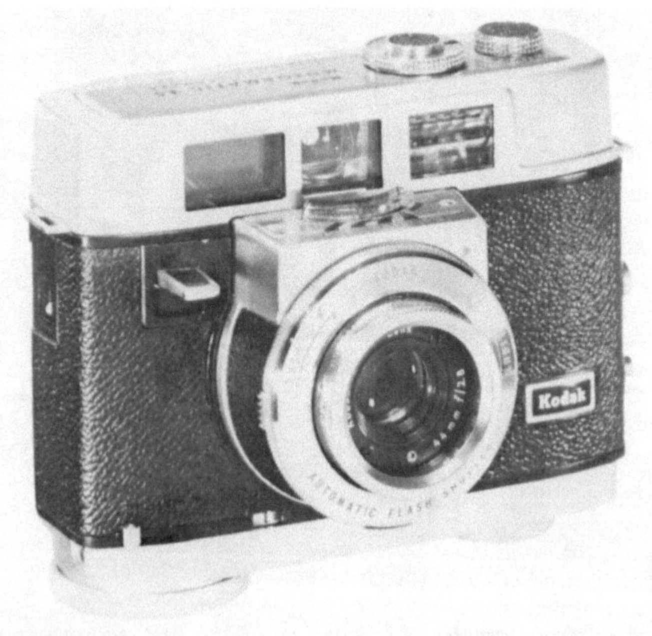

A deluxe model of the Kodak Automatic 35 was this Motormatic 35. A spring-wound motor automatically advanced the film as each exposure was made. To Kodak's pleasant surprise, it was a best seller.

tive when seen attached to the camera, it was awkward to hold and use and in some models not particularly dependable. Designs like the Kodak Rotary Flashholder, which held six M2 bulbs in a plastic mat that could be rotated between shots (which eventually led to Sylvania's flashcube), failed to generate much enthusiasm with the public. But then, one might reasonably ask, "What cared Kodak?" It had contributed several fine, classic designs to the field; the introduction of Kodachrome film had assured the 35mm format of continued success at a time when the experts were seriously concerned about its survival; and today Kodak still outsells other film manufacturers by a wide margin. Since the Kodak Instamatic camera now rules the roost (see Chapter 18), the company sells all the cameras it can mass-produce. Who could ask for anything more?

12 But Everything Was Perfex

One of the better known of the small manufacturers which came into existence as a direct result of the 35mm craze during the latter thirties was Candid Camera Corporation of America. Candid Camera differed from the others in that its Perfex line was successful enough to perpetuate itself through a variety of models. Incorporated in May, 1938, by Carl Price, Benjamin Edelman, and Joseph Price, Candid Camera occupied the fourth floor of a six-story brick building at 844 West Adams Street in Chicago.

The first Perfex was a most unusual-looking Bakelite box which bore a faint and offhand resemblance to the early Argus line. It was designed by Leonard W. Gacki and Joseph Price and was announced to the trade in October, 1938, as the 35mm Speed Candid. Utilizing a focal-plane shutter (1/25 to 1/500 and *B*), which operated behind an *f*/3.5 Graf-Perfex anastigmat lens in an interchangeable helical focusing mount, the Perfex had a certain amount of shock appeal which attracted attention to its user. It was large (6½ × 3½ × 1¾ inches), rather clumsy to use, and weighed over 2 pounds—it was no lightweight in the 35mm field.

The camera front was a metal plate riveted to the plastic

body and carried the body shutter release above and left of the lens. A telescopic viewfinder of molded plastic at the top far right was an integral part of the camera body. Left of the viewfinder was the shutter selector knob; a metal housing containing the narrow-base, noncoupled rangefinder, the exposure counter, the film-release button, and the winding knob. Also incorporated into the body was an extinction meter centered on the bottom directly beneath the lens, between the film-rewind knob and the tripod socket.[1]

Access to the interior was gained by pushing outward on spring-loaded clamps at each end of the back and removing it completely. The back was made of base metal, and the inside was covered with an imitation felt-fleck paint to prevent reflections. The designers' initials (LWG–JP) were cast into both the back and the camera interior. The pressure plate was not precision-machined but simply a piece of inexpensive metal beveled slightly at each end to allow the film to pass through without catching. Between the spring-loaded clamps on the outside of the back was positioned a small circular exposure dial for use with the extinction exposure meter.

All these features were available in this striking 35mm for a mere $25; additional lenses, a 2-inch *f*/2.8, a 4-inch *f*/4.5, and a 6-inch *f*/4.5 telephoto lens could also be purchased. An accessory interchangeable ground-glass focusing back was also made available for use with a set of four lens-extension tubes.

This model was first sold to a catalog house (Triangle Electric) and was subsequently distributed through photo-

[1] Unlike later Perfex models (and contrary to generally accepted practice), the film traveled from right to left.

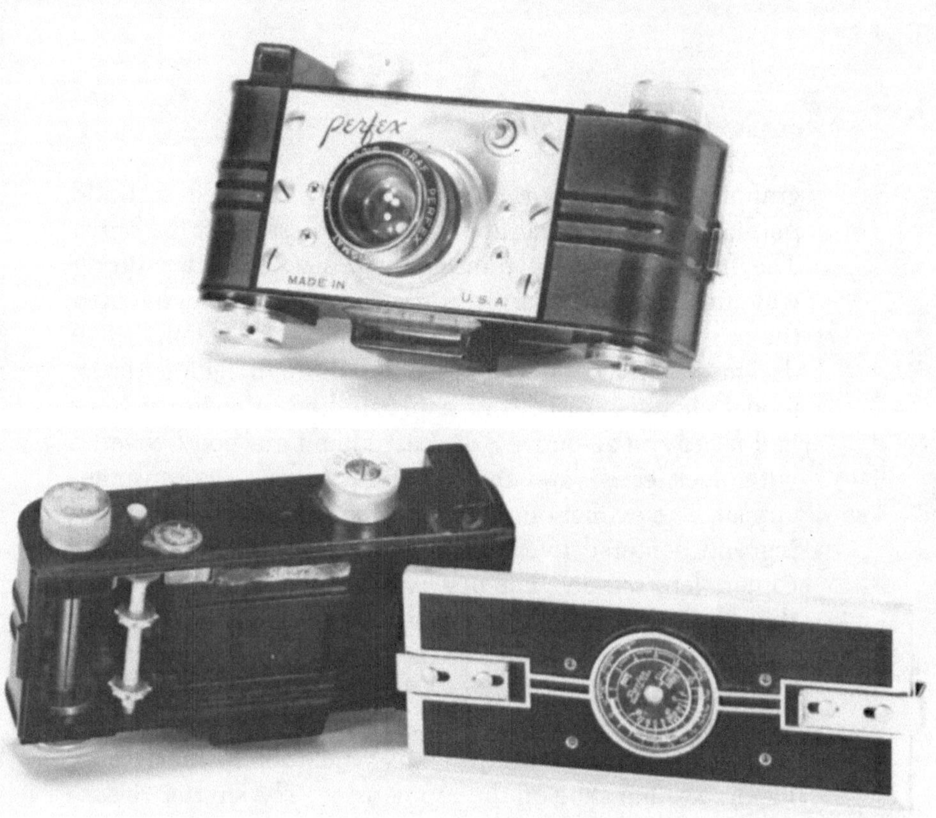

The original Perfex was neither a thing of beauty nor a joy to behold, but it worked and somehow sold. In many respects it resembled a king-sized Argus A, but its focal-plane shutter (the first made in an American 35mm camera) made it unique. The camera shown here is missing the noncoupled built-on rangefinder, which was attached between the shutter-wind knob and the film-wind knob on the top. The shutter release is at the left of the lens, and the shutter is wound with a knob on the right-hand side, making the camera truly two-handed. At the bottom center of the camera is an extinction exposure meter. Introduced in 1938, this first Perfex was produced for only a short time; the company soon switched from plastic to a die-cast metal body with a coupled rangefinder which was to be standard through several model changes.

graphic shops, with advertising mainly confined to trade publications. But it was apparently successful just the same. The following year a thoroughly professional face-lifting was announced, the Perfex Forty Four. This model retained the original Perfex concept but was styled in a competitive design which would remain basic through half a dozen model changes. Instead of a plastic body the Forty Four had a heavy (22-ounce) die-cast aluminum body covered with black leather and chrome trim to increase its consumer appeal. The extinction meter was moved into the top housing and centered over the lens to "read light just as the camera lens sees it." The new wide-base rangefinder was of the separate-window (one tinted green) and split-image variety, but now coupled to the lens. The Forty Four was available with either an *f*/3.5 ($37.50) or an *f*/2.8 ($47.50) Graf-Perfex anastigmat lens. The full-range focal-plane shutter (1 to 1/1250 and *B*) was synchronized for flash by means of a hot shoe on the top housing. The shutter release was positioned within the center of the shutter selector dial, and on the right face of the camera were four small nubs and two swinging levers; one lever controlled the direction of film movement (forward or rewind), the other was a fast-slow shutter selector. As with its predecessor, the Perfex Forty Four had a distinctive appearance, but, while the new design was more contemporary and attractive, it belied the camera's awkward balance, clumsy handling, and poorly machined components. The parts were almost certainly subcontracted for manufacture, with Candid Camera assembling them, the procedure also used by Universal Camera Corporation at the time.

The Perfex Forty Four fitted into the hand better than its predecessor but proved difficult to use efficiently. A large

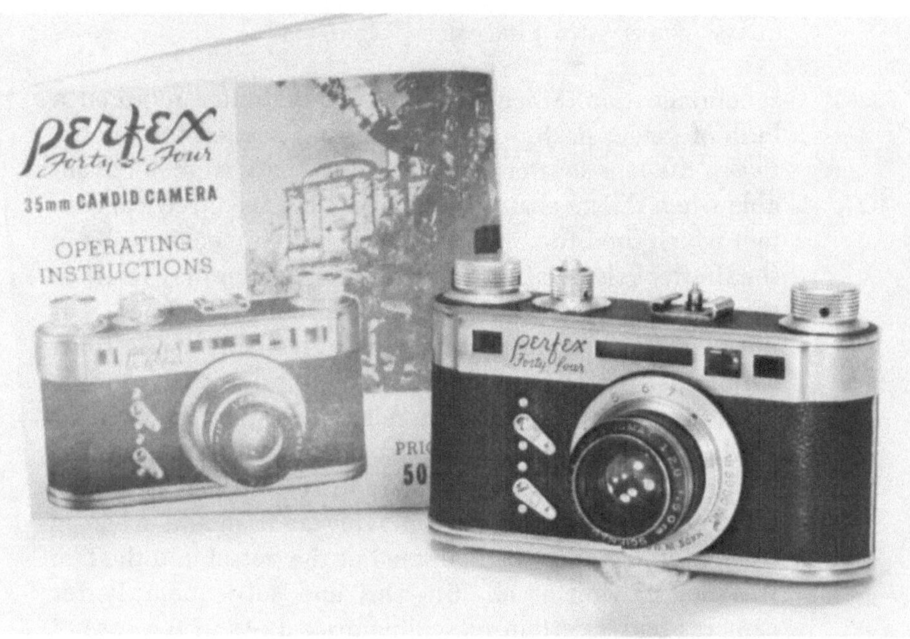

Die-cast of aluminum, the Perfex Forty Four marked the beginning of a line of distinctively styled 35mm cameras which Candid Camera Corporation introduced after its first plastic model. If the shutter speeds and other letters and numerals had been stamped with more finesse, the camera would have had a quality look. A previous owner added an ASA flash terminal to the hot shoe of the model shown above.

milled ring with engraved footage markings which formed the rear of the lens mount was revolved to focus the lens (a slow procedure by any standard) and had a tendency after some use to loosen to the point where setting the lens aperture after focusing changed the focus. While the camera took acceptable pictures, its shutter was especially prone to difficulties, mainly in the slow-speed range and in the flash synchronization. Since the focal-plane shutter required a bulb of longer flash duration than that needed for the between-the-lens shutter, and since such bulbs were not available when the mechanism was designed, the electrical contact which fired the bulb was incorporated mechanically in the shutter-release assembly and did not accurately relate to the position of the shutter curtain. In short, the camera was not really synchronized for the flash available at the time; variations in the flash firing were introduced depending upon how rapidly the shutter release was depressed.

The coupled film transport also acted up on occasion, and it was not unusual for the camera to act and feel as though a full 36-exposure roll were moving as it should when in reality the transport had jammed at the outset and the film was not moving at all. But this and subsequent Perfex cameras had a certain masculine appeal. As in the case of the Argus C-3 (but to a lesser extent), they seemed to answer the call for rugged, outdoor cameras, and they enjoyed a certain popularity which was almost certainly helped by aggressive marketing on the part of the national distributors and the catalog houses which handled them. Thus the Perfex succeeded where other and better 35mm miniatures like the Clarus MS-35, its closest competitor in appearance, failed (see Chapter 16).

Two more Perfex models were announced in 1940, the Thirty Three and the Fifty Five. The Fifty Five was almost identical to the Forty Four in exterior design, except that the lever switches on the front had become round buttons. The top button (for rewind) was now a push button and much smaller than the other, which had a stationary center around which a ring with engraved shutter speeds revolved. Using an *f*/3.5 or *f*/2.8 Scienar lens manufactured by General Scientific Company, of Chicago (which also made the Graf lenses), the Fifty Five was priced $2 above the Forty Four. It was the only Perfex known to the authors to show design evolution within a single model designation. The Fifty Five first appeared in an all-chrome version with an extinction meter. As World War II approached, the film-wind, rewind, and shutter selector knobs were finished in black. The Fifty Five reappeared in 1945 as an all-chrome model without the meter and equipped with a Wollensak lens.

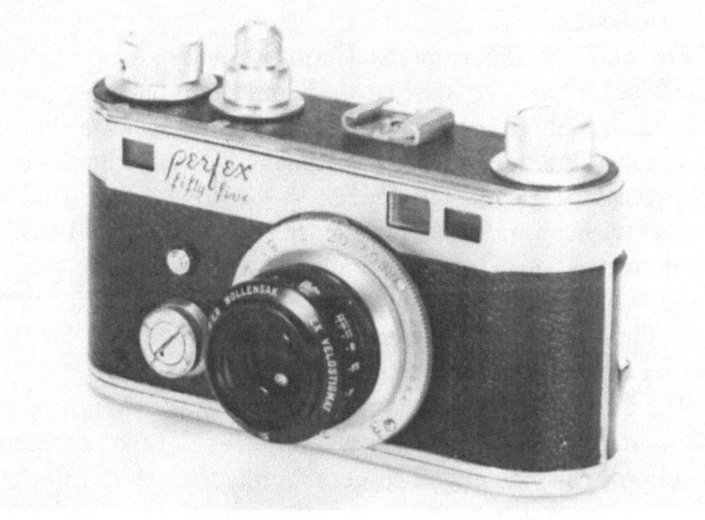

With a die-cast body and a better-quality Wollensak lens, the Perfex Fifty Five returned to the market after the war as a good seller, and for a camera of the period, a good picture maker. Although the shutter acquired a bad reputation and the finish was somewhat crude, it had a loyal band of users. The prewar Fifty Five had been produced with a built-in extinction meter.

An economy version of the Fifty Five, the Thirty Three was finished mainly in black and featured an *f*/3.5 Scienar lens in a limited-speed focal-plane shutter with speeds of 1/25 to 1/500. The face of the camera had only a control for rewind; no fast-slow switch was necessary.

In 1942 the Thirty Three was replaced by the Perfex Twenty Two ($44.50). Again it was basically the Fifty Five body with the *f*/3.5 Scienar lens. The Fifty Five now had a Wollensak Velostigmat *f*/3.5 ($57.50) or *f*/2.8 ($69.50) lens. The increase in price was indicative of both rising costs and the better-quality lens, but apparently few Fifty Five's were marketed.

For each of the cameras Candid Camera Corporation furnished a carrying case, a flash unit, slip-on filters, and a 6-inch telephoto lens made by General Scientific—a simple task, for all models after the original one were identical in all external respects. The Perfex Vu-or-Project was a table viewer with a 6-inch ground-glass screen which could easily be converted to a projector, giving a rather weak 5-foot image when placed 7½ feet from the screen; its price was $25. Those were all the accessories made available for the Perfex cameras by their manufacturer.

The Perfex Fifty Five reappeared briefly after World War II. In 1947 it was replaced by the Perfex deLuxe, a heavy and improperly balanced camera. While the exterior design was very similar to that of the Fifty Five in shape and size, the deLuxe had a stamped body and an *f*/2.8 ($99.50) or *f*/2.0 ($175.00) Wollensak Velostigmat lens. The body contour was accentuated slightly in the stamping process, with a sort of outward flange where the top housing met the body. The exposure counter, shutter selector dial, and accessory clip were not set into the housing top as in previous models, but they retained the undersized and poorly engraved numerals. The extinction meter had disappeared, but the split-image rangefinder remained the same, dim and difficult to use. The lens, with three sets of *f*-stop markings engraved in white around the black barrel, was removable, and even the focusing mount could be removed by depressing the small plunger (which indicated the distance setting while preventing rotation beyond either footage limit) and continuing to unscrew it.

The bottom of the camera was freed by rotating a circular disk from the closed to the open position. Caution in removal was required, for the take-up spool was not attached to the wind knob and usually fell out when the back was separated from the body. The purpose of this procedure was twofold: for cassette-to-cassette loading or to allow the user to attach the film leader firmly to the spool and drop it into place, a rapid-loading system found on several 35mm's of the period. The interior paint was poorly applied, with runs and flecks alternating with tiny bare spots, and the chromed film path and the pressure plate were poorly machined and easily scratched.[2]

While earlier Perfex shutters had developed difficulties in the slow-speed range, the deLuxe had a tendency to lose speeds above 1/50. The shutter-cocking gears moved into

[2] Lest the reader gain the impression that there were no redeeming qualities about the Perfex line, the authors go on record with praise for an interesting 8mm magazine movie camera that made its appearance in 1947 along with the still-camera line. The camera had a turret front holding three lenses and featuring five operating speeds from 12 to 64 fps. The spring-wind body had all the attributes of a Bendix washer of the era, including the rattling coffee-grinder sound, but the camera delivered one of the sharpest and steadiest pictures of any magazine movie camera in its price range ($69.50 with an *f*/2.5 lens; $105.90 with an *f*/1.9 lens), and no amount of careless handling affected its operation. It was a real workhorse.

position when the film was wound but often slipped, carrying only the capping portion of the shutter curtain across the film plane. The user was unaware of this problem until his cartridge of film was developed and returned with no exposures on it. Earlier Perfex shutters had suffered through the years from shutter lag—slowing down before reaching the end of the frame—and so did the deLuxe shutter.

Exterior finish was little better than interior finish. The grained-leatherette covering was bady fitted, paint disappeared from the engraved lettering almost immediately, and the cutout in the top housing to allow the focusing scale to seat properly was not covered from the inside, a flaw that allowed dust and dirt to enter and clog the shutter and other interior mechanical works.

The same body also appeared as models One-O-One and One-O-Two (minus the focal-plane shutter and selector knobs), bearing a Wollensak *f*/4.5 ($39.50) or *f*/3.5 ($49.50) lens in a Wollensak Synchro-Alphax shutter. It would be interesting to unravel the internal affairs of the Camera Corporation of America. In 1950 many of these bodies found their way to various camera outlets, fitted with uncoated Kodak Ektar lenses mounted in Compur Rapid shutters and sold in the $50 to $65 price range. Construction and general appearance remained shoddy; one of the authors (Lahue) bought a Perfex One-O-One with an *f*/3.5 Ektar lens through the mail. Upon receipt he was convinced that he had been sold a used camera. The interior was poorly machined, badly scratched, and even burred in places along the film path—apparently common flaws in Perfex construction. Subsequent investigation confirmed that it was indeed a new Perfex and that the lens-shutter assembly was one of a large number manufactured before the war in Germany

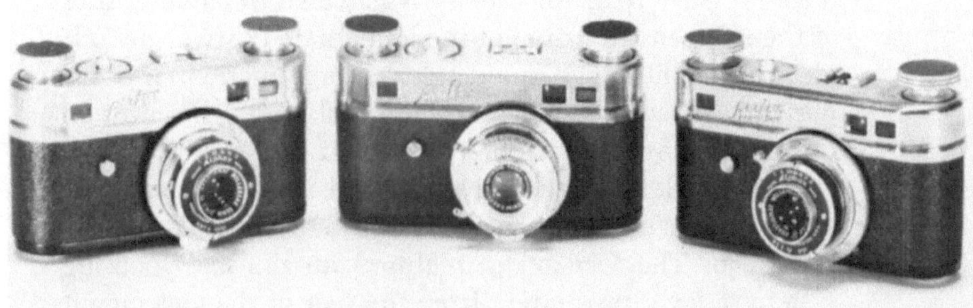

The last of the traditional Perfex models were the One-O-One (f/4.5, left) and the One-O-Two (f/3.5) series. In these models the standard Perfex body was gutted of its focal-plane shutter, and a front shutter substituted. Both models used the Wollensak Alphax lens-shutter combination. A fortunate turn of events occurred when a group of surplus prewar Kodak Retina lenses and shutters, uncoated and unsynchronized, turned up in Germany. They were imported by an independent company and mounted on the Perfex One-O-One and also on the One-O-Two (right). Perfex sold the bodies but did not mount the lenses. The camera thus produced was beyond any doubt the best picture taker ever made under the Perfex marque.

for use by Kodak in its Retina models and seized by the United States Army during World War II.

The single departure in Perfex design came at the close of its existence. Named the Cee-ay 35, this camera had an unusual history. Designed by Leonard W. Gacki and introduced in early 1949, the Cee-ay 35 had all the characteristics of the inexpensive 35mm of the immediate postwar period: a coupled split-image rangefinder, an automatic film transport, an exposure counter, and built-in flash synchronization through the Synchro-Alphax shutter. It was available with either a Wollensak *f*/4.5 lens (1/25 to 1/150, *T* and *B*) at $49.50 or a Wollensak *f*/3.5 lens (1/10 to 1/200, *T* and *B*) at $59.50. The Cee-ay 35 featured an unusual focusing system: a lever protruding from the rear of the lens mount was operated by the user's left forefinger, and directly opposite, on the other side of the mount, were the footage scale (3 feet to infinity) and an indicator which was connected to the focusing lever.

Camera fans were somewhat taken aback when Ciro Cameras (manufacturer of the popular twin-lens Ciro-flex) announced the Ciro 35 in September, 1949, only eight months after the Cee-ay 35 appeared—it was the same camera at the same price. Ciro had apparently decided that the easiest way to expand into the 35mm line was to purchase a design complete with dies. While the Perfex line died, the Cee-ay 35 lived on under its new name. When the Camera Corporation of America ceased operation, its assets were sold to Ciro, including tools and dies for all its cameras. Unfortunately, no records have survived, and it is impossible to reconstruct or even accurately estimate the number of cameras manufactured in each Perfex model line. Even the serial numbers bear no relationship among models.

Ciro made but three minor exterior alterations in the Cee-ay 35. The film-wind knob now carried two sets of small millings instead of one large set; the faceplate on either side of the lens mount was converted from black on chrome to all chrome, with a black line and numerals breaking up the pattern; and the new name was engraved where the old one had appeared. But the story does not end there

Ciro was facing a difficult period. Its major offering was a line of roll-film reflex cameras, and the twin-lens-reflex vogue that rose after World War II (and produced the Argoflex, the Kodak Reflex I and II, the Ansco Reflex, and several foreign designs) was gradually nearing an end.

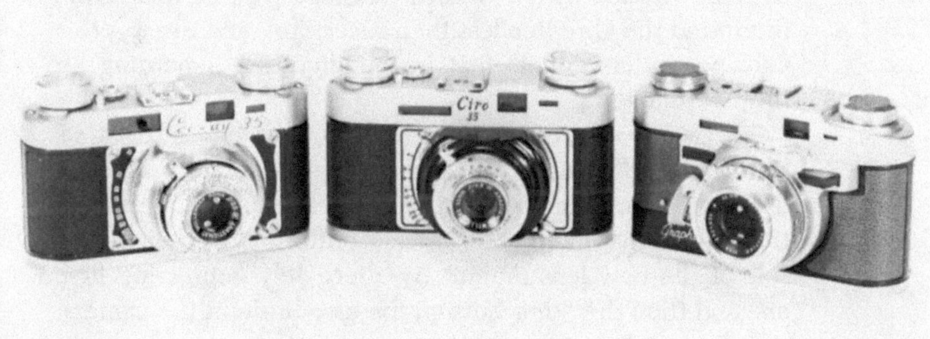

The only American camera made by three different manufacturers was the Cee-ay 35. The camera was designed and produced by Perfex (left). Perfex later sold its camera business to Ciro, which produced the camera with slight modifications as the Ciro 35 (center). Ciro in turn sold out to Graflex, which continued to manufacture the camera unchanged as the Graphic 35. The trim was later modified, as shown at the right, but the camera remained essentially the same model.

Ciro's entry into 35mm was an attempt at broadening its market, but it was unable to weather the competitive storm. In 1951, Graflex, Inc., which had distributed the Ciro 35 since its introduction, acquired the entire Ciro line. The Ciro 35 became a Graflex camera, again with a minimum of changes.

Three models were marketed: the R, the S, and the T.[3] At first the only alteration was in the lens mount, which became black instead of chrome, and Graflex marketed the camera with the *f*/3.5 Wollensak lens in a Synchro-Alphax shutter and the words Graflex, Inc., engraved on the base plate. A chrome lens mount identified the model carrying a lever-set Century shutter (1/10 to 1/200, *T* and *B*) as the Synchro-Alphax with a Graflex Graftar lens. But, alas, Graflex found the Ciro line less than successful, and in 1955 the Ciro 35 underwent its first major alteration, emerging as the Graphic 35.[4]

The result was a somewhat different camera, but not so startlingly different as to be unrecognizable from its predecessor. More sophisticated in appearance and features, the Graphic 35 made use of rocker buttons placed on either side of the new lens mount. By alternately depressing first one and then the other button, the user brought the camera

[3] The Ciro models R and S were introduced in September, 1949. The R was discontinued in August, 1952; the S, in September, 1954. The Ciro T appeared in August, 1951, and was discontinued in September, 1953.

[4] The Graphic 35 appeared in five different models, priced from $77.50 to $87.50, from February, 1955, to July, 1957. They can be most easily identified by the type of Spectromatic scale used on them. Each model was introduced and discontinued on the following dates: with *f*/3.5 lens: single color band (February, 1955–April, 1956); double color band (April, 1956–May, 1957); universal band (April, 1956–July, 1957). With *f*/2.8 lens: double color band (November, 1955–April, 1956); universal band (April, 1956–July, 1957). See also note 5 below.

into focus by means of the separate window rangefinder. Because of lower manufacturing costs abroad, the *f*/3.5 and *f*/2.8 Graflex Graftar lenses were in reality of German (Rodenstock) manufacture. They were placed in Prontor SVS rim-set shutters of the double-action (cock-before-release) type. Several types of these shutters appeared in the Graphic 35. It was entirely possible, indeed probable, that a camera offered to the customer in one shop would carry the traditional shutter markings of 1, 1/2, 1/5, 1/10, 1/25, 1/50, 1/100, and 1/300 second, while down the street another shop carried Graphic 35's with the European designations 1, 1/2, 1/4, 1/8, 1/15, 1/30, 1/60, 1/125, and 1/300 second.

All functional settings were visible from above on these new Graphic 35's, and a three-way calculator, built into the top of the lens-mount housing between the focusing buttons, contained a depth-of-field scale, a quick-set footage scale, and a Spectromatic flash-setting color code.[5] One final feature wrapped up the Graphic 35 in a neat package: the body shutter release. The release, an upside-down comma on the right face of the camera below the focusing but-

[5] This quick-set flash calculator came in three types: single, double, and universal bands. The single band was factory-calculated for Kodachrome and a No. 5B or Press 25B bulb. Below each aperture number was a different color band. Focusing as for daylight pictures, the user set the aperture indicator to the color band matching the one appearing on the focusing scale. The double-band model had two rows of colors beneath the aperture numbers and allowed the use of the calculator with more than one emulsion. The universal band had an additional setting on the underside of the lens mount which allowed the scale to be preset for any film-bulb combination with a guide number from 30 to 300. This handy little combination was no doubt offered with the best of intentions, but it served mainly to confound the already confused owner. Its real value was problematical, for after the instruction book explained its foolproof operation, the owner was cautioned to remember to adjust the aperture according to the lightness or darkness of his surroundings.

ton, was designed to be tripped with an outward motion of the middle right finger. Those accustomed to pushing down or in on the shutter-release button were now asked (for the sake of steady pictures) to swing their finger in careless abandon to do the same job. How many Graphic 35's were discarded as a result has not been recorded for posterity.

The major design features previously mentioned were retained in succeeding models (the Graphic 35 Electric and the Jet even incorporated a built-in meter), but production was transferred to Japan, and subsequent cameras are by definition beyond the scope of this book. Thus several years and many variations later, the last of the Perfex cameras designed by Gacki disappeared from the marketplace. Of all the American 35mm miniature cameras, the Cee-ay 35 undoubtedly had the most unusual heritage, appearing under three different marques with virtually no change.

And what happened to the Camera Corporation of America? When its assets were sold to Ciro, Joseph Price purchased the remaining inventory of Perfex cameras, its repair rights, and the company name. Price helped Ciro begin manufacture of the Cee-ay under its own name, serving as an engineering consultant. Once servicing of the Perfex had been stabilized, Price and his associates, Ed Broderick and Morris Litwin, provided in- and out-of-warranty service in the Middle West to various camera manufacturers who wanted to compete with the strong repair facility network Revere Camera Company had established in the heartland of the country. Thus International Camera Corporation came into being. It was established at a time when there was a need for its services, and with its engineering and manufacturing background International was able to express product difficulties in terms easily understood by manufacturers. It was also able to make emergency parts as well as redesign a product for foreign manufacture. Today International is the largest and best-known repair service company in the world.

13 Except for Teachers on Vacation with Their Bolseys

For almost a decade after World War II the Bolsey 35mm was a common sight on photo dealers' shelves. Today it graces the windows of countless pawnshops, and occasionally one turns up in the used-camera department of a photo shop. During its prime the Bolsey was probably the only 35mm miniature favored to any large degree by women. Although the Bolsey sold well, its dainty size made it a difficult camera for male hands to hold and use easily, and it found its largest market among career and professional women.[1]

The story of this ladies' favorite dates back to the early twenties, when Jacques Bolsey designed the Cinegraph Bol, the first combination still and movie camera which used 35mm film and could also be used as a projector for both formats. After producing this camera in his native Switzerland from 1923 to 1926, Bolsey introduced the Bolex 16mm movie camera in 1927, a design he produced until the autumn of 1931, when Paillard & Cie, a Swiss firm, bought the manufacturing rights. Bolsey moved to the United States

[1] When one of the authors (Lahue) was a high-school student in the late forties, every woman teacher in the school's English department owned a Bolsey B, as did several of the elementary teachers in the community.

in 1939 and was involved in a number of projects during the war years, including work for the United States government.

Publicly chastising the American photographic industry in 1947 for its high profit margin and contending that American cameras were needlessly expensive, Bolsey introduced his first American-made camera, the Bolsey B. He had concluded that he could capture a sizable portion of the market with a low-cost, high-quality 35mm camera. Industry reaction was a knowing smile, for, innovative camera designer that he was, Jacques Bolsey had yet to learn about doing business in the American economy. Only five years later, Bolsey would find himself backed against the financial wall because he had neglected to take into consideration one factor his competition understood only too well. Within five years labor costs had doubled owing to fringe benefits, wage increases, and the ever-increasing complexity of labor regulations. Since the original Bolsey B was marketed well below the industry price for similarly equipped cameras, its profit margin had shrunk ominously.

The Bolsey B was an extremely compact double-frame 35mm camera featuring a coated 44mm *f*/3.2 Wollensak anastigmat lens set in an Alphax shutter (1/10 to 1/200, *T* and *B*).[2] Weighing but 14 ounces, the camera's aluminum die-cast body was attractively covered with leatherette and a polished-chrome trim. A coupled split-image rangefinder focused the helical mount from 2 feet to infinity by means of a lever on the lens mount operated by the left finger. A cor-

[2] While 44mm was actually the mathematically correct focal length for the double-frame negative, Bolsey concentrated his advertising on the slogan that the "special 44mm lens increases depth of focus." Bolsey *was* shrewd enough not to try to reeducate the American public as regards lens design.

responding depth-of-field calculator graced the camera back. But the Bolsey's chief sales point was its extremely compact size (2¾ × 4 inches), which meant that very little film was wasted in loading the camera. Because of the short distance between the supply and take-up spools, the owner could get up to four additional pictures beyond the stated number on a 36-exposure load, a film economy unmatched by any other 35mm camera on the market. Turning the lock lever on the bottom allowed the entire back to slide off. The film was then loaded in the usual manner and the cover replaced and locked. Film was advanced by lifting the wind knob and turning slightly in the direction of the arrow before releasing the knob and continuing to wind until the mechanism locked. The film advance proved to be one of the camera's less charming features, and some owners found it a little difficult to use.

An external micro synchronizer was designed for use with the Bolsey B.[3] A hollow screw held a small plastic attachment to the cable-release socket, which was placed just below the shutter-release lever. Two prongs extending below the unit served as electrical connections, and a flat contact spring on top of the attachment tripped the flash when the release lever touched it. The micro synchronizer was designed to be left on the camera permanently.

The Bolsey B2 added internal flash synchronization through a Synchro-Alphax shutter by means of a contact plug and screw on the flash unit, which were attached to corresponding holes on the camera back just behind the re-

[3] The Bolsey Flash Gun No. 1 was provided by Camera Optic Company, a leading flashgun manufacturer of the era and one of the first to promote the battery-condenser (BC) concept, which was an industry vogue for a few years until inexpensive electronic flashguns from Japan took over the market.

wind knob.[4] The B2 also had a rather ingenious double-exposure-prevention device, which was brought into operation by advancing the film; a spring-loaded piston placed near the release lever moved outward from the body as the shutter-release lever was depressed.[5] This movement was accompanied by an audible click, notifying the user that he had passed the point of no return. Further pressure tripped the shutter, and the piston now retarded a second exposure by blocking the release lever's return to a preset position. Winding the film retracted the piston, allowing the lever to return to a cocked position. The system could be bypassed for intentional double exposures by simply depressing the piston by hand, allowing the shutter-release lever to override it and recock.

Jacques Bolsey continued to exercise his ample design ingenuity, coming forth with another model in 1950—the Bolsey C—during the twin-lens-reflex vogue. In essence the C was the original B design adapted to a reflex mode, but with a plus factor: it was the only low-cost 35mm reflex on the market at the time. With a Bolsey C the user had his choice of the regular Bolsey eye-level viewfinder and rangefinder or true reflex viewing and focusing through a viewing lens placed above the objective lens and between the rangefinder windows. When the rangefinder was actuated, the C provided a full and unreversed ground-glass image. It also incorporated a flip-up magnifier inside the folding reflex hood, just as its larger relatives did. The addition of flash

4 A removable-lens version of the B was combined with a special ring flash unit (electronic) for medical and scientific work and was designated the Bolsey B Special.

5 The Bolsey shutter release had a lever travel of ½ inch, an interminable length for a slow, steady tripping motion.

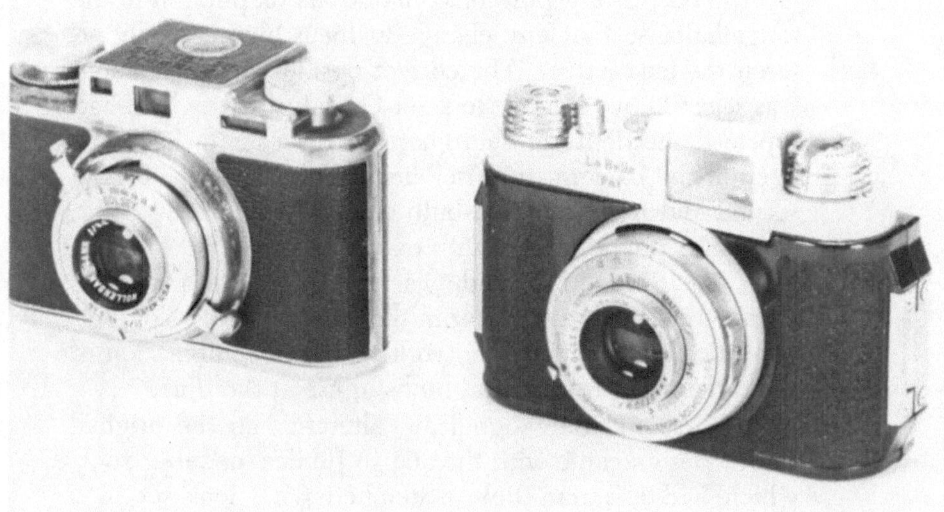

The original Bolsey Model B (left) was a good buy with its fast lens, film economy, and built-in rangefinder. But the film advance left much to be desired, and many owners could not operate it. The La Belle Pal (right), originally appearing as the Bolsey Model A in 1953, is a camera of doubtful parentage. La Belle denied making it, and the company has no records of its production. All-plastic, with an f/4.5 lens, three-speed shutter, and a low price, it was a good buy, but it met with not-too-eager reception.

synchronization to the C resulted in the C2, and the Set-O-Matic flash system introduced in 1953 on the B22 and the C22 models was a further—and final—refinement.

Advertised as an automatic-diaphragm flash computer, the Set-O-Matic consisted of a slide underneath the lens mount which could be adjusted to engage the aperture lever in one of four positions, linking the diaphragm to the rangefinder so that any change in focus automatically altered the lens setting. The correct position of engagement was selected by referring to a Set-O-Matic chart which had replaced the depth-of-field chart on the camera back. To operate the camera, the user first set the shutter at 1/50 second and found his film-bulb combination on the chart. The chart provided him with one of four letters: A, B, C, or D. The Set-O-Matic slide was then lifted and moved to engage the diaphragm lever in the corresponding slot. Set-O-Matic was calculated to work with any combination of the three popular films and bulbs in use at the time.[6]

The first and only significant alteration in the original Bolsey design came with the $69.50 Jubilee model in 1955, which had a 45mm Bolsey-Steinheil *f*/2.8 lens set in a Bolsey-Gauthier synchroshutter (1/10 to 1/200 and *B*). The release was moved to the top of the viewfinder housing. A new ratio-gear system was incorporated in the film-wind mechanism, and double-exposure prevention was now incorporated internally. The Set-O-Matic system was updated to include the new, faster color emulsions and expanded to daylight conditions with light and subject conditions engraved on ring scales in the lens mount. The chosen shutter speed was matched to the film in use, and the diaphragm ring was revolved until the light condition matched the type of subject. In effect, this daylight use of Set-O-Matic was simply a sophisticated translation of the information which each film manufacturer included with his product.

But none of these innovations stemmed the basic prob-

[6] This flash calibration system was relatively simple to use and its effectiveness in ease of use was far superior to the color-coded system which Graflex used on its Graphic 35 (see Chapter 12).

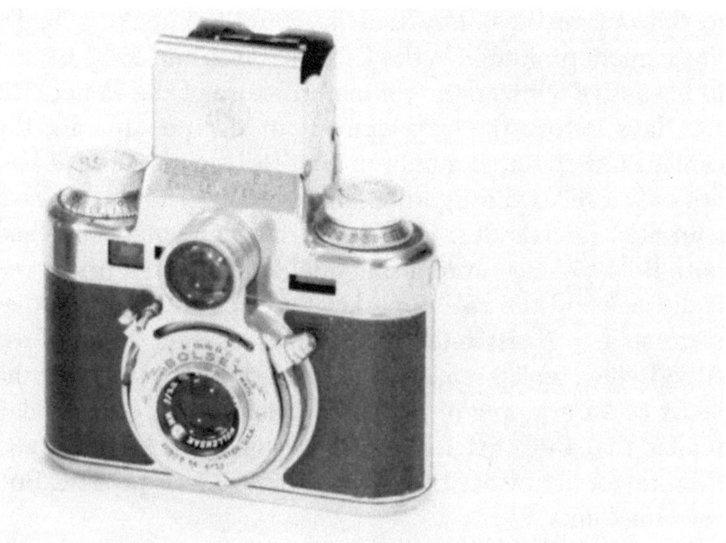

The Bolsey Model C combined the best of two worlds, as Bolsey had previously done with the original Bolsey Reflex (now the Alpa). The Model C was both a twin-lens-reflex and a rangefinder camera. Actually little more than the Model B2 with another lens and viewfinder, it was a novel idea but either a bit ahead of its time or not quite what the public wanted. It never became a major item in the Bolsey line.

lem faced by Bolsey—rising costs. Three other less-expensive models bore the Bolsey name: the Explorer (*f*/2.8 lens and rapid-wind lever), the Bolseyflex (a 120 roll-film reflex with *f*/7.7 lens and flash synchronization), and the Pal (a simplified version of the B without rangefinder, distributed by La Belle).

Of all Bolsey's designs perhaps the most brilliant was his "ultimate" camera, the Bolsey Reflex. It was obvious even to the optimistic Bolsey that he could not make a really fine camera profitably in the United States, and so he turned to his native Switzerland for manufacture of the camera. In the days before the development of the pentaprism, the camera incorporated a unique dual focusing system: it had not only a reflex through-the-lens viewing system but also a split-field rangefinder. Unfortunately, the camera also had both Bolsey's uncompromising quality and a high price, and it did not find the ready market that he foresaw. The dies were sold to Swiss interests, and the camera became the Alpa Reflex, which remained in limited production over the years as an example of Swiss precision. Few people today realize that the Alpa Reflex was designed in America, and even fewer are aware that it was a product of Jacques Bolsey's ingenuity.

When foreign competition and rising costs finally forced Bolsey out of the still-camera business in 1956, he designed a unique movie camera using a standard single-8mm film (like the UniveX) and small enough to be carried in a shirt pocket. Its most unusual features were its adjustable shutter speeds and capability for use as a small still camera. While the Bolsey Single 8 never gained much popularity with the public, it was sold in substantial quantities to government intelligence agencies both in America and abroad. Its main drawbacks were that it used a cassette that only Bolsey supplied and that, while it was a true precision instrument, its price was high for a movie camera. In a cost-cutting move manufacture of the camera was also moved to Switzerland, but even with an economy model (the Uniset 8), the move was too late; Kodak introduced Super 8mm. Jacques Bolsey died in January, 1962, and his widow and son assumed control of what remained of another American dream.

14 An Imitation Failed . . .

The second American attempt to produce a sophisticated precision 35mm miniature camera was the Kardon, introduced by Premier Instrument Corporation of New York City in early 1947. By no means an original design as the Ektra had been, the Kardon was a virtual copy of the Leica IIIA. How it came into being makes a somewhat unusual story.

With the tide of battle turning in favor of the Allies in 1944, E. Leitz, Inc., the American distributor of the Leica, contracted to produce Leica cameras for the United States Signal Corps. A lengthy search to find a subcontractor able and willing to undertake the project proved near-futile. No one wanted to attempt the manufacture of the famed camera, for its specifications were high and tolerances extremely close. Hearing of the search, Peter Kardon of Premier decided that the challenge was interesting and potentially profitable. It is said that patriotism also played a role in Kardon's decision. He felt that the ingenuity and capability of American industry should not default in the face of a set of German specifications, which, having been reached once before, could certainly be reached again. Whether or not professional pride and patriotic fervor actually influenced

Kardon's decision that much, he did publicly vow to make a Leica that would not only meet the specifications but exceed them and result in a *better* Leica.

Kardon was awarded the contract and set to work at once, completing his first camera just as the war ended. Leitz's subcontract with Premier ended in September, 1945, soon after V-J Day and Kardon, who now had a substantial investment in camera dies and machinery, decided to continue manufacturing the camera. Since Leitz owned the rights to the Leica name, Kardon gave the camera his own name and arranged for distribution through post exchanges and later through retail stores.

While the challenge Kardon had accepted was a stiff one, he managed to duplicate the Leica in appearance and function, achieving the close tolerances required and yet allowing for complete interchangeability of all 452 component parts for easy repair and replacement. That in itself was a major accomplishment, for Leica parts were bench-fitted during assembly. Three major changes were made in design and production: the Kardon body was manufactured of die-stamped Duralumin, the lens mount was noncollapsible, and built-in flash synchronization was incorporated in the civilian model marketed in 1947.[1] Except for these alterations, the Kardon remained a Leica IIIA in disguise, at a price to consumers of $262.50 plus tax.[2]

When Premier contracted to develop and produce this "American Leica," Leitz lenses were unavailable, and the Kardon was designed for use with the 47mm *f*/2.0 soft-

[1] The Kardon lens mount extended only 5/16 inch farther beyond the body than the original Leica mount in its collapsed position.

[2] The Kardon initially sold at $393.50, tax included. By 1948 the price had been reduced to $262.50 plus 10 per cent tax—$288.75.

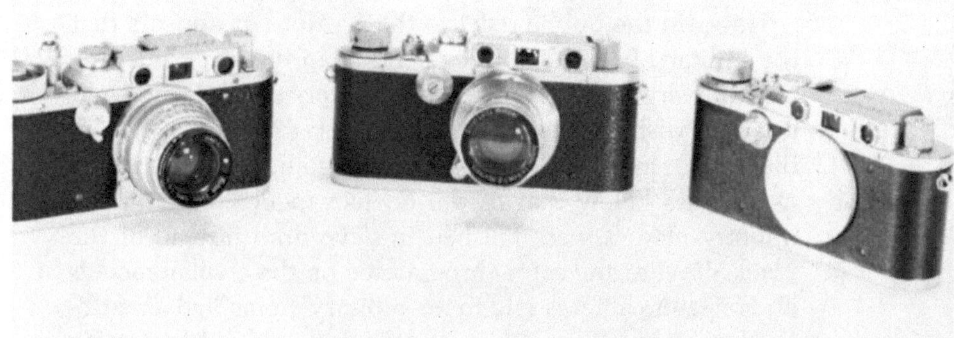

A direct copy of the Leica IIIA (center), the Kardon was made in two versions: one for the military (left) and one for the civilian market (right). Note the large winding knob and the extended shutter release on the military version. The lens shown is the Kodak Ektar f/2.0 with its unique wheel focusing mount. On early versions of the Kardon the wheel was placed directly over the slow-shutter-speed knob, but the position was later changed to that shown on the camera at the left. On the camera at the right the lens has been replaced by a screw-in body cap.

coated Ektar lens (originally manufactured for the Kodak Bantam Special), which Kodak made available.[3] Although the helical focusing mount rotated through 504 degrees to assure critical focusing with ease, the position of the focusing wheel was poorly chosen, and while the user was adjusting it, his finger was likely to nudge the slow-speed dial, inadvertently dislodging it from its setting. The owners'

[3] The *f*/2.0 Ektar lens was also made available for use on the Leica. Sold separately at $120 plus tax, it coupled with the Leica rangefinder and was supposed to "bring the camera up to date."

objections to this small but annoying flaw were immediately forthcoming, and it was soon corrected in production.

How did the public react to the Kardon? It appears that the military bought more cameras than civilians did. The military version had an extended shutter-release button and a larger, dish-shaped film-winding knob (about 1 inch in diameter), and the focus wheel was mounted at a position 45 degrees below that of the civilian model. Most of the military models were finished in olive drab instead of the black Vinylite and satin chrome used on the civilian models and on later cameras sold to the military. Some had a matte-finish metal plate on the camera back to be used for jotting down notes. Most consumer sales were made through post exchanges, and only 5,000 Kardons, give or take a few, were manufactured. The camera proved a failure from a sales standpoint, but down through the years its reputation has suffered somewhat more than necessary for this disappointing market showing.

There were reasons for the Kardon's rejection by the public, and not all were sound, rational ones. It was made from an honored German design by an American manufacturer, and therefore it could not be as good as the Leica, or so the reasoning went. The snob appeal was lacking. By the consumers' yardstick it cost too much for a mere copy, and reappearance of the "real" Leica, whose price soon became competitive, placed the Kardon at a disadvantage. Its appearance suffered from the use of the austere Vinylite instead of grained leather, and the camera also lacked that indefinable feel of quality which was expected in its price range. Kardon's was not a unique case; early Canon cameras were also copies of the Leica, and they suffered from the same reaction. But the engineering was a feat of which

Peter Kardon had every right to be proud, although Kardons which the authors have used seem to lack the silky smoothness of operation characteristic of the Leica IIIA. This was especially true of the focusing mount. The action was stiff and tended to stiffen still more with age unless alcohol was applied periodically.

As a combat photographer in Korea during the police action of 1950 to 1953, one of the authors (Lahue) was issued the traditional Speed Graphic, a Bolsey B used by his predecessor, and a brand-new Kardon, still in its factory packing. The Kardon was a joy to behold, sparkling and gleaming as it caught the pale October sunlight rippling across its smooth chrome finish. It was not olive drab and thus was apparently one of the later Kardons sold to the army. There could be no faulting the optics; the Ektar was crisp and biting-sharp in definition. But the camera body handled and functioned on a level not too far above that of the Russian Fed which the author had earlier acquired.[4] His reaction to the camera was not as violent as that of less-tolerant members of the Signal Corps fraternity of photographers who came in contact with the Kardon, but a badly machined pressure plate made the camera impossible to use without severely scratching the film as it was transported. Since replacement pressure plates were unavailable in Korea, the local Signal Corps repair depot in Seoul refused to accept it for repair.[5] That particular Kardon spent its

[4] The Fed was a Soviet copy of the Leica manufactured after World War II.

[5] In the Leica tradition only the Kardon bottom came off for film loading, and visual inspection of the pressure plate and film path was impossible without extensive knowledge, tools, and time. Time was something Signal Corps photographers seldom had in Korea. Repair parts were not available in rear-echelon depots in Japan.

time on active duty perched on a wooden shelf above an army cot in the battalion photographers' tent.

What has happened to the Kardon in the twenty-odd years since its introduction? Rumors have it that the United States Navy has all the dies packed away in one of its storage warehouses, conveniently forgotten. Why the navy has them no one has satisfactorily explained to the authors. Premier has long been out of business. Occasionally a Kardon pops up in the larger photo stores for sale, usually at a highly inflated price, for dealers have a suspicion that it is a collector's item. As this is being written, a Kardon bearing the serial number 1286 sits beside the typewriter, in near-mint condition for its age. The satin-chrome finish is still like new, and the spring sunlight dances across its housing as it did so many years before. While its shutter still has the voice of authority as it travels across the film plane, the lens mount

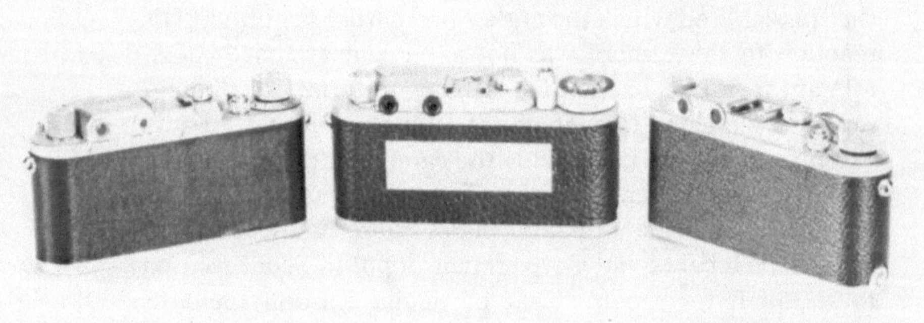

Rear views of the Kardons and the Leica IIIA. The military version of the Kardon is distinguished by the writing plate on the back of the body. Can you distinguish the civilian model from the Leica? (Note differences in the eyepiece corrections and sizes of the openings.)

is as difficult to focus as ever, and the somnolent Ektar looks to the future, which holds little for it to do.

Questions about the camera are still unanswered. Why was the Leica IIIA chosen as the prototype when the IIIC was in production in Germany? Why could not a patriotic appeal to an established camera manufacturer, such as Argus, Graflex, or Kodak, have produced results if a real need for such a camera existed? For that matter, the Ektra could have more easily been reactivated at a fraction of the cost. Why was the Leica chosen over the Contax? The questions and the speculation seem endless.

15 But So Did Originality

Photography's brave new world, which had begun with the Ektra before World War II, came of age in early 1949 with the entry of Bell & Howell into the still-camera field with a deluxe 35mm miniature, the Foton. It was introduced at the staggeringly deluxe price of $700, which fortunately included the federal excise tax.[1] A case, a flash attachment, and other accessories were available at additional cost. Bell & Howell claimed that a million dollars and ten years of engineering had gone into the development of the Foton, making it the classiest 35mm since the Ektra. But Kodak's Ektra has become something of a legend among 35mm fans, while the Foton is all but forgotten.

When the Foton was introduced, Bell & Howell's advertising material stressed that it would not be mass-produced but would be precision-crafted through every stage of its manufacture. To Bell & Howell this meant that supply would not equal demand for some time; but to the company's great chagrin demand never really caught up with production.[2] The Foton quickly became a financial night-

[1] By November, 1949, the list price had been reduced to $498, where it remained throughout the Foton's short life.

[2] At the time of this writing, Henry's Camera in Los Angeles still had a demonstrator model for sale at $285.

mare; it was the photographic industry's Edsel. Bell & Howell still prefers to gloss over the camera in discussing its past lines, contending simply that the Foton was "ahead of its time." But no effort has ever been made to revamp and reintroduce the camera or any variation of it.

The camera boasted such features as an all-metal focal-plane shutter (1 to 1/1000 and *B*) on a single control dial, a coupled coincidence rangefinder incorporating a dichromatic reflector beam splitter, and built-in flash synchronization through a standard-size accessory clip. It was guaranteed for life against defects in workmanship or materials.[3] Where the camera departed from the usual and entered the realm of the exotic was in its spring-driven film transport, which was capable of six pictures a minute. Until the advent of the Kodak Motormatic a few years later the Foton was the only American-made 35mm sequence camera.

The film transport was wound with a semicircular ring on the bottom left of the camera, which folded into the body contour when not in use and was snapped down 90 degrees to form a sort of winding knob. Winding the transport not only was a bit difficult because of the small quasi knob but was also time-consuming. The shutter-release button was placed on the camera face, across the lens from the speed-selector dial, and by manipulating a ring-and-pin control the user could set it to give either a single or a multiple exposure or lock the shutter. A touch of class was added with a sunken exposure counter enclosed in a metal casing with a glasslike top. Looking very much like a watch dial, it had a red needle arrow which rotated with each exposure.

But the feature which Bell & Howell expected would put

[3] The shutter-control dial, a knob similar to the one on the Universal Mercury, was positioned in a like manner on the face of the camera.

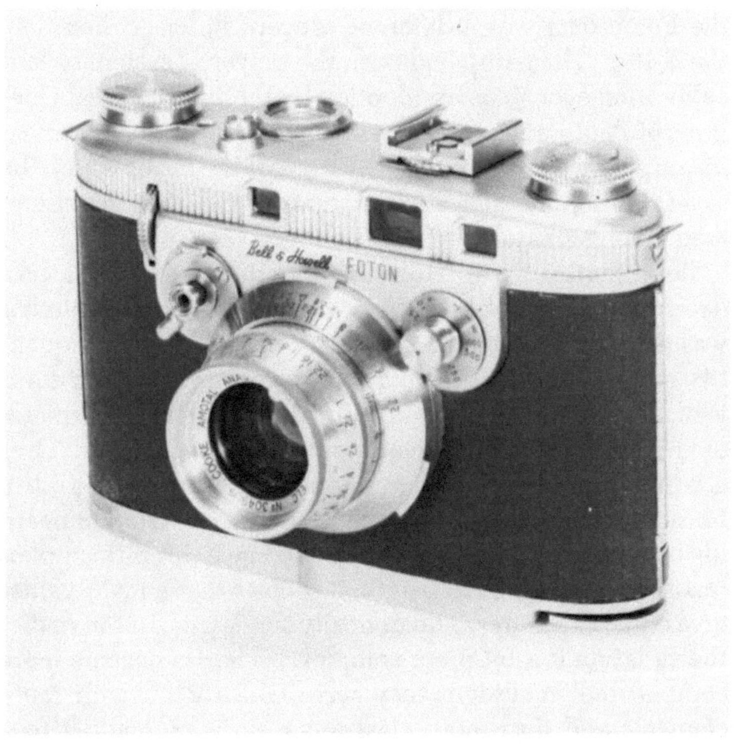

The Bell & Howell Foton was to be the standard by which all other 35mm cameras could be measured, but unfortunately its manufacturing cost was far greater than its sales. Within a year the list price was lowered to almost half the original, but even then it did not move from dealers' shelves. In this camera Bell & Howell attempted to introduce the T-stop to amateur photography (it had been used in movie photography for years), but the public could see no use for it. The camera used a focusing wheel similar to the Contax and took bursts of pictures, eliminating the need for a rapid winder. There are persistent reports that new Fotons are still available.

the Foton over was a different concept in lens calibration, the T-stop. The *f*-stop had been the universal system of lens calibration ever since its adoption by the International Congress of Photography in Paris in 1900. The *f*-stop had served its purpose well in bringing order out of the chaos which the several different calibration systems of the 1890's had created among amateurs.

Representing an aperture opening which allowed a predetermined amount of light to pass through, an *f*-number was nothing more than an arbitrary designation for a quantity of light reaching the film. The determination of *f*-stops is made by dividing the lens's focal length by the diameter of the aperture at any given point in its adjustment. Thus a 2-inch lens with a maximum opening of 1 inch is said to have a speed of *f*/2.0; the same lens with a maximum opening of ½ inch has a rated speed of *f*/4. Since this ratio applies to any lens of any focal length, it follows that equal *f*-values give equal exposures. Theoretically this is true, but in reality the situation is a bit more complex. As lenses became more complicated in design, they were constructed with more elements and thus more glass-to-air surfaces, leading to a degree of light loss during transmission. Lens coating helped reduce this loss. Optical engineers calculated that an uncoated lens lost approximately 4 per cent light and that coating reduced the loss to only 1 per cent. The value of coating can be shown mathematically in a lens with four air-spaced elements, giving eight glass-to-air surfaces. Such a lens would transmit 24 per cent more light than the same lens without coating.

But the average camera purchaser was unaware of these details of optical engineering, and as long as his camera produced acceptable pictures, he could not have been less concerned about the technical aspects of lens calibration, a fact which the American Standards Association (ASA) and the Society of Motion Picture Engineers (SMPE) failed to take into consideration in 1948 when they devised the new T-stop system, in which the T represented the actual light transmitted. T-stops were determined slightly differently from *f*-stops: the lens's focal length was divided by the diameter of an open hole which transmitted light in an amount equal to the aperture under calibration.

The English firm producing Taylor-Taylor-Cooke lenses embraced the new system and, in conjunction with Bell & Howell, devised a T/2.2 coated Amotel lens (equal to an *f*/2.0) for the Foton.[4] Born with the Foton, the T-stop system also died with it when the factory run of cameras gradually sold out and the line was discontinued, leaving thousands of T-T-C lenses without bodies. The lenses were eventually sold to Peerless Camera, of New York City, which placed them in Leica mounts and sold them on every camera body they could find, especially the Nicca.[5] Peerless not only marketed the lenses in its own store but also wholesaled them to any and every company willing to buy them. Strangely enough, despite all the T-T-C lenses that were sold, finding a used one today in a camera store or pawnshop is a virtual impossibility, akin to finding a Kardon Ektar *f*/2.0 lens in a Leica mount, which appeared a year or so before the Foton.

While the Foton was a lightweight but bulky and greatly

[4] The Nicca was a Japanese copy of the Leica sold by Sears, Roebuck as the Tower 35mm and later purchased by Yashica.

[5] Interestingly enough, while the Amotel's diaphragm was calculated in T-stops, the front of the lens mount carried engraved lettering identifying it as an Amotel anastigmat 2 inch *f*/2 ELC, in addition to the serial number.

overpriced camera, it fell neatly into the hand. Positioned perfectly, the shutter release was depressed inward with a mere squeeze, lessening the possibility of camera movement. Slightly above and to one side, the rangefinder-adjustment wheel was a knurled vertical disk which protruded from the body and revolved easily. Once the distance was set, the finger dropped naturally to the shutter release to make the exposure. But the price of the camera, its unusual appearance, and the unfamiliar T-stop system combined to create a negative impression on the part of would-be purchasers, who agreed that it was quite a hunk of camera, but not for them. The crash course in educating the public to lens calibration failed miserably, and Fotons graced the windows and shelves of camera stores for years after Bell & Howell discontinued manufacture. It was an engineering success but a commercial failure. The last vestiges of the camera noted by the authors were large numbers of Foton flash attachments being sold over a bargain counter of the Camera Barn, a New York photographic distressed-merchandise house, in 1969. A few months later they were gone, and with them the brave new world of 35mm photography.

16 Has-Beens and Never-Weres

In addition to the cameras already discussed, many 35mm miniatures appeared and disappeared with regularity. In the thirties and again in the fifties it seemed that nearly everyone considered himself a camera designer. With a bit of financing and a dash of talent companies were formed overnight, and while some of the resulting cameras faded slowly from the marketplace, many disappeared overnight.[1] While advertising and distribution were factors as important as the cameras themselves in these failures, some of the 35mm miniatures characterized an era and as such are worthy of brief examination.

Perhaps the outstanding over-the-counter failure was the Clarus MS-35. A heavy, solidly built 35mm which bore an off-the-cuff resemblance to the Perfex series (could the same design hand have figured in both?), it was announced in 1939 by International Photographic Industries, Inc., of Chi-

[1] The Cand-O-Matic camera, produced by Millfred Manufacturing Company, of Los Angeles, is a good example. It was introduced in the spring of 1938 at $49.50. The metal body, resembling that of the Argus C-3, was equipped with a Wollensak *f*/3.5 fixed-focus lens and a leaf shutter (1/25 to 1/100, *T* and *B*). Its outstanding feature was a tension-spring film wind which allowed the user to take 15 consecutive pictures in 5 seconds.

cago. It was the creation of Paul Mann and was to be available for sale in May, 1940, on a franchise basis. Equipped with an interchangeable but uncoated lens made by Busch Precision Camera Corporation, the Clarus was to be marketed with either an *f*/3.5 or an *f*/2.8 anastigmat lens in a collapsible focusing mount.[2] Its other announced features were by then standard in 35mm cameras: a built-in coupled split-image rangefinder, a separate eye-level viewfinder with parallax adjustment, an automatic film transport and frame counter, an accessory shoe, and a focal-plane shutter with speeds to 1/1000 second. Accessory telephoto and wide-angle lenses were to be made available, with prices starting at $59.50 for the camera equipped with the *f*/3.5 lens.

But the war interceded, and while the Clarus MS-35 reached prototype stage, it was not actually produced until 1946. In that year it appeared, virtually unchanged in appearance from its prewar design. Encased in a machined body, the camera's intermittent-curtain focal-plane shutter was made of rubberized nylon. Built-in flash synchronization was added later. The camera was available with either an *f*/2.8 Wollensak Velostigmat lens ($116.25) or an *f*/2.0 Wollensak Raptar lens ($168.50), and interchangeable 35mm *f*/3.5 wide-angle and 101mm *f*/3.5 telephoto lenses were provided, with accessory viewfinders for each.

The camera was manufactured by the Clarus Camera Manufacturing Company, of Minneapolis, owned by Paul Mann, Nate Owen, and the law firm of Robbins, Lyons and Davis. Mann contributed his design and the tooling pro-

[2] The Busch lens was made by the manufacturer of the Busch Press camera, the major competitor of Speed Graphic, but one that never caused Graflex executives to lose a good night's sleep.

duced before the war, and Owen set up and managed the plant with financing arranged by the attorneys. As Mathew Broms, Clarus' last general manager, assessed the company for the authors, Clarus represented "a magnificent opportunity for great success slipping through the fingers of a group of grossly inexperienced, poorly financed people, saddled with a camera whose basic design was a complete failure—in the beginning."

To meet the fantastic consumer demand for cameras after

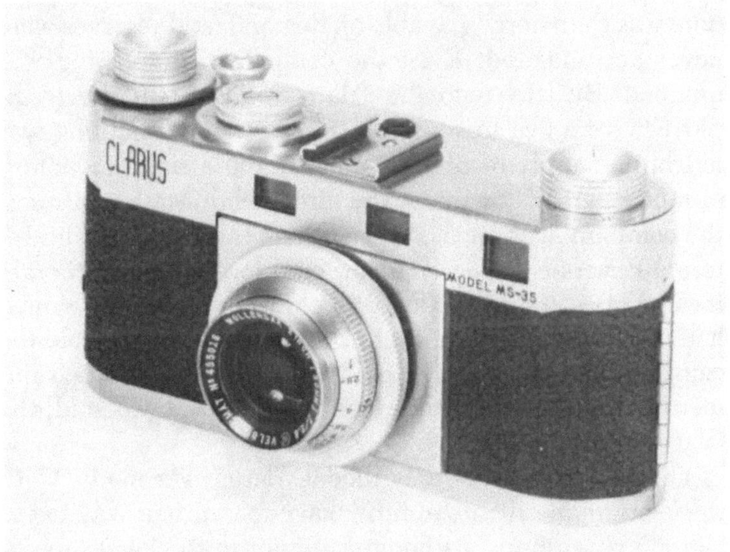

The Clarus MS-35 was the only model made by the Clarus Company. Announced before the war, it was produced afterward by a group of camera manufacturing innocents. The camera was poorly made and worked erratically almost until the end, when the maker finally learned how to correct the flaws. By then it was too late, and the MS-35 became another name of the past.

the war, Clarus rushed the MS-35 into production. The design was so bad that mechanical failures plagued over 20,000 cameras produced between 1946 and 1948. Most of the problems were finally corrected, and by 1949 Clarus finally had a good camera. But it was a case of too little, too late. Consumers and in particular Clarus dealers had soured on the camera, and with many other good cameras available, why should they buy a box reputed to be a lemon?

From a financial standpoint, Clarus was in trouble from the beginning. No long-term financing was available, every note was short-term, payable on demand, and reserves were never accumulated. Even the company's advertising was minimal. By late 1950 the salaries of the remaining employees were five to six weeks in arrears, and nothing was left but a supply of unsold cameras and a small inventory of repair parts. Now under the direction of Mathew Broms, the company was on the verge of bankruptcy, with the Internal Revenue Service a major creditor—not for corporate income taxes but for taxes withheld from employees which had not been turned over to the government. Broms sold the camera and its accessories to La Belle Industries in the summer of 1952, and once the reserve of cameras was sold, the Clarus MS-35 died.[3]

The MS-35 was the only model Clarus ever made. Over the years an inordinate number have found their way to the nation's pawnshops. Although superior to the Perfex in de-

[3] La Belle Industries (now La Belle Sales Corporation, Oconomowoc, Wisconsin) entered the retail field in 1946 with an automatic slide projector developed during the war to teach aircraft identification. Its growth was rapid and consistent until competition arose in the automatic-projector field. La Belle experienced difficulties in maintaining its share of the market, which had not grown as rapidly as the number of competitors (21). It acquired the Clarus and the Pal (originally a Bolsey) for diversification; both were losers, and La Belle gave up on cameras.

sign, manufacture, and finish, the Clarus was not as easy to use as its appearance and price would lead one to believe. It was probably the most disproportioned 35mm ever made. Its length and height were no problem, but the body from front to back was far too wide for comfortable handling, and proper balance seemed to be lacking in spite of its solid feel. Unfortunately, when the user held a Clarus in the manner most comfortable to him, his left index finger automatically fitted over the left rangefinder window, rendering it useless and necessitating a shift of the camera in the hand. To make matters (and operation) worse, focusing was accomplished (as with the Perfex) by revolving a milled ring at the extreme rear of the lens mount, a slow and awkward procedure appropriate to photographing still-life and tabletop subjects rather than action pictures.

The Clarus' rangefinder and viewfinder were placed in separate windows at the rear of the camera—a common feature of cameras of the period—but the Clarus reversed the normal state of affairs by placing the rangefinder window at the right of the viewfinder, making it necessary for the user to shift his eye from right to left before framing his subject (natural eye movement is from left to right). While the shutter was accurate through the range from 1/20 to 1/500, not even the company would bet on what happened when the 1/1000 setting was used. Otherwise the mechanical design of late production runs appears to have been sound enough, but the large number of mint-condition MS-35's available today (all with the *f*/2.8 lens) seems to invite the question just how much use owners gave their cameras.

An equally interesting but more unusual design appeared in 1946 as the Vokar I. Originally announced for production

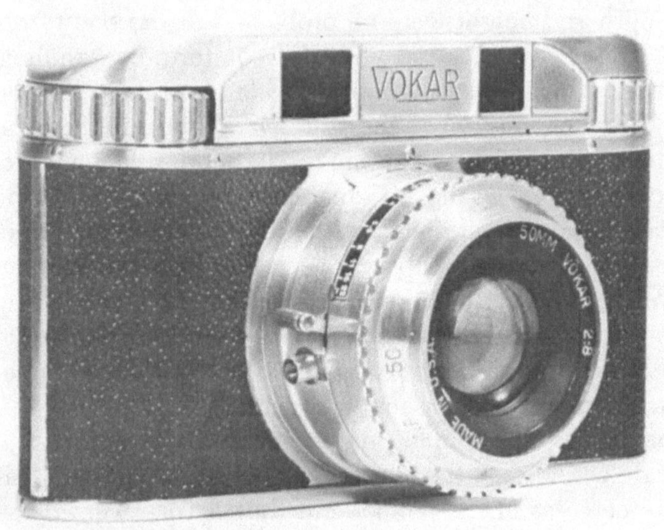

The Vokar was made by the man who founded Argus, and the Vokar plant was but a few miles from Argus' doorstep. Designed by Richard Bills just before the war, the Vokar was unique, compact, and potentially a good seller. The company produced all the components of the camera, including the lens and the shutter, in contrast to other small manufacturers of the period who generally subcontracted for components and served mainly as assemblers. A subsequent Vokar II was identical except for the nameplate. The camera's failure in the marketplace was apparently the result of several factors, primarily lack of marketing muscle. Vokar made more money producing inexpensive slide projectors for Sears, Roebuck.

in 1940 by the Electronic Products Manufacturing Corporation, of Ann Arbor, Michigan, the Vokar seems not to have been actually introduced until 1946. The Vokar Corporation was owned and operated by the Verschoor Corporation, which the founder of Argus had established in the late thirties. The designer of the camera was Richard H. Bills.

Priced at $76.50 plus tax, the streamlined, aluminum-bodied Vokar I was ahead of its time both in appearance and in operation.[4] The Vokar plant in Celina, Ohio, produced the coated *f*/2.8 anastigmat lens and the shutter, but many of the camera's parts were subcontracted for manufacture and brought together for assembly in Dexter, Michigan. Bills had designed a unitary body with film-wind and rewind knobs as integral parts of the camera and not the protruding appendages common to 35mm-camera design. The coupled rangefinder was incorporated in the viewfinder window, and all adjustments and controls were visible from the top of the camera. The leaf shutter was full-range (1 to 1/300 second) but not synchronized for flash.

With its large lens mount, which appeared somewhat out of proportion to the rest of the camera, the Vokar was a bit ungainly in appearance, but it fit the hand nicely and operated smoothly. Owners who had used other cameras found it difficult to become accustomed to the integrated film-wind knob, which necessitated a series of partial turns to wind an exposure into place. But therein lay another interesting design. The behind-the-lens shutter had a double-action movement, requiring cocking before use. Where

[4] For all practical purposes the Vokar I never made it to market; a later Vokar II was indistinguishable in appearance and features. The Vokar cameras cost about $250,000 to put into production, and just about broke even.

virtually every other double-action leaf-shutter design incorporated in the 35mm miniatures of the period required the user to set the shutter manually before releasing it, in the Vokar shutter cocking was coupled to the film transport. Near the top of the take-up spindle in the film chamber was a small metal plug extending from the shaft. After making an exposure, the user began winding the film-advance knob, which also activated the internal shutter-cocking mechanism. After one complete revolution of the film-wind knob, as the shutter was nearing its fully cocked position, a small pin lever dropped into place to catch the metal plug as it revolved a second time, preventing further winding. Tripping the shutter moved the pin lever out of position, freeing the film-wind transport for another cycle.

The Vokar line also included two inexpensive slide projectors, one of metal and one of Bakelite, as well as enlarging accessories and a refraction liquid (Scratch Patch), which could be applied to negatives to render scratches invisible during printing.[5] At its peak the company employed 200 people, but camera production was discontinued in 1947, and the Vokar Corporation gradually faded from view, along with the Vokar I and II. Ineffective marketing techniques, a camera that never clicked with the public, and too few salesmen to promote the cameras reduced Vokar to bankruptcy. The remaining stock of cameras and equipment which could not be sold was scrapped. The buildings were disposed of, and the operation disappeared into that

[5] The Scratch Patch was also developed by Dick Bills, who had used it at Argus many years before. Bills mixed it at home and sold it to the company in gallon quantities, which the company rebottled. Vokar did not want to buy the formula. Years later Bills sold the formula to Edwal Laboratories, which promptly marketed it under the name Edwal No-Scratch.

twilight zone of one-time camera manufacturers whose dreams exceeded the grandeur of their product.

Little success greeted the introduction of the Zephyr candid camera in the autumn of 1938. A product of Photographic Industries of America, in New York City, the Zephyr was handled by three of the nation's leading photographic distributors,[6] and while on paper its features and specifications were well worth the trade price, the camera disappeared during the war and remains one of those intriguing mysteries which plague researchers.

The Zephyr's hard aluminum-alloy case weighed 16½ ounces unloaded, was covered with leather and a brushed-aluminum trim, and housed a focal-plane shutter with a speed range of 1/25 to 1/500 and *B*. An interchangeable Wollensak lens set in a helical focusing mount was the Zephyr's eye on the world, and the camera could be purchased with either an *f*/3.5 ($22.50) or an *f*/2.9 ($29.50) Velostigmat lens, both stopping down to *f*/16. A small box on the top housed a Wollensak telescopic viewfinder and an accessory clip.

The film ran through the camera from right to left (in the manner of the first Perfex), and the film-wind knob and frame counter were at the extreme left in the position usually reserved for the rewind knob. The shutter-speed-selector dial rested under a matching knob, placed about halfway between the viewfinder and the extreme right of the camera, giving it a slightly awkward appearance; yet this design provided a positive grip and gave the Zephyr a good handling balance. Except for the placement of the body

[6] The distributors were Raygram Corporation in the East, Hornstein Photo Sales in the Middle West, and Seeman's, Inc., on the West Coast.

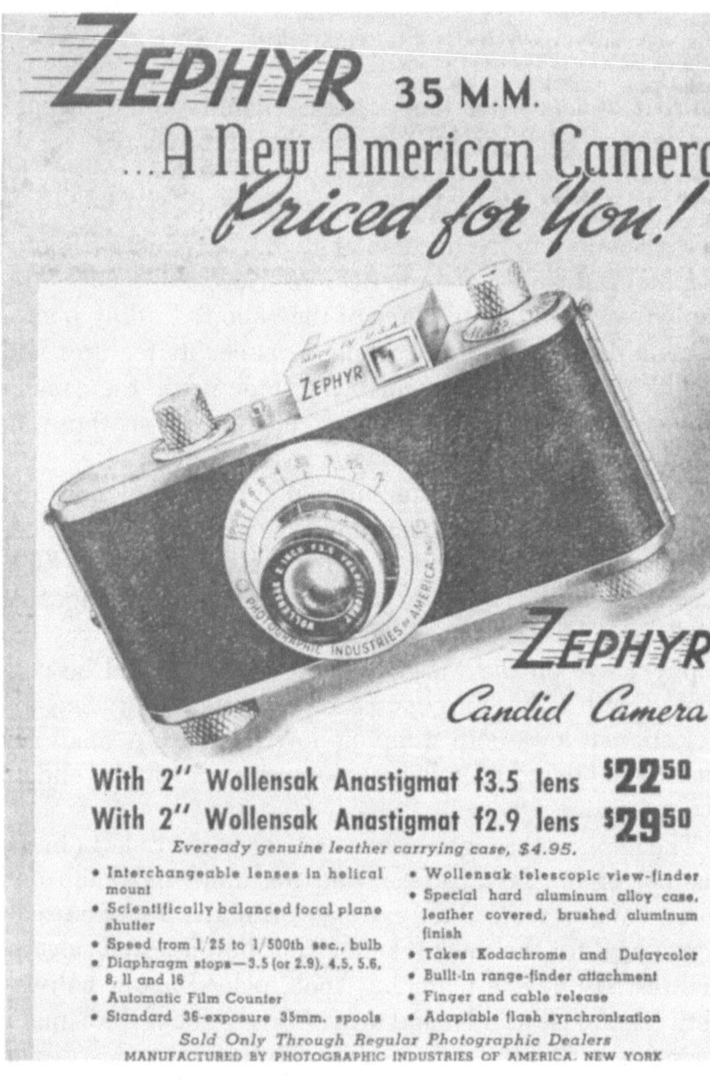

The Zephyr 35mm appeared and disappeared with amazing rapidity. While most of the American 35mm cameras can be found in pawnshops across the nation, the Zephyr is as scarce today as the Detrola 400.

shutter release midway between the shutter-speed selector and the viewfinder, the design was uncluttered and refreshingly simple.[7] A milled knob was positioned on each end of the Zephyr's base; one rewound the film, and the other contained a tripod socket. The Zephyr's main problem seems to have been its focal-plane shutter, which had the annoying habit of binding. But for $22.50 what could one expect, a Leica?

Even more interesting—and obscure—was the Detrola 400, a camera manufactured and distributed by the Photographic Division of the Detrola Corporation, of Detroit, Michigan.[8] The 400 was Detrola's first and only 35mm; previous Detrola cameras, such as the D and E, introduced in 1939, used 127 roll film. They were roughly equivalent to the Argus line in design and construction and were virtually indistinguishable from the Falcon series (see below). The 400 was introduced in the summer of 1940 and appears to have been the first American attempt to duplicate the Leica while avoiding charges of patent and design infringement. For this reason a rather detailed discussion of this camera is worthwhile.

Reviews in the photographic magazines openly hinted that the Detrola 400 was America's own Leica, and indeed it bore an interesting resemblance to its German counterpart, both in features and in appearance.[9] The all-metal

[7] The placement of the shutter release introduced a certain amount of difficulty. It was hard to trip the shutter without interfering with the selector's revolution. To counteract the problem it was necessary to alter the potentially good grasp of the camera.

[8] Detrola's major product was the Pee-Wee radio.

[9] The Leica's external dimensions with lens extended were 5½ × 2⅝ × 2½ inches. The Detrola's dimensions with noncollapsible lens were 5½ × 2¾ × 3⅞. The weights of the cameras, unloaded, were 20½ and 23 ounces, respectively.

stamped body was covered with black leather, with satin-chrome finish on all exposed metal parts. The back was unlocked with a 45-degree swing of the recessed base lever and slid down and off the body. The design of the top housing was strikingly similar to that of the Leica but was functionally different and slightly more streamlined. While both cameras used narrow-based, separate-window superimposed rangefinders, there were several important alterations in the Detrola's controls.

At the left of the camera (from the user's viewpoint) was the film-rewind knob, a split key or disk similar to that used on the Argus Colorcamera. The rewind control was on the camera face, at the right of the lens. An accessory shoe for attaching a flashgun or optional viewfinders to be made available for accessory lenses was placed slightly off center above the rangefinder and viewfinder windows. At the right were two disk turn buttons which governed the focal-plane shutter. By lifting and rotating the larger of the two buttons, the user could adjust the normal-speed range from 1/25 to 1/1500 and *B*.[10] For slower speeds the larger dial was set on 1/25 and the smaller, nonlift disk revolved to one of five marked positions, 1/10, 1/5, 1/2, 1/ and OUT. The OUT position was the neutral setting used when the faster speeds on the other dial were to be used.[11] A body plunger release completed the top housing controls, with a film-wind knob and exposure counter sitting on the camera body proper.

[10] The range of speeds was quite different from that of the Leica. The 400 shutter had markings of 1/25, 1/50, 1/75, 1/100, 1/200, 1/500, and 1/1500. The 400's main problem was its shutter. A right-to-left horizontal-travel single-adjustable slit-cloth curtain, it was a free-wheeling adaptation of the Leica shutter and slowed noticeably near the end of travel.

[11] Some cameras were marked with an S instead of OUT to indicate the neutral position.

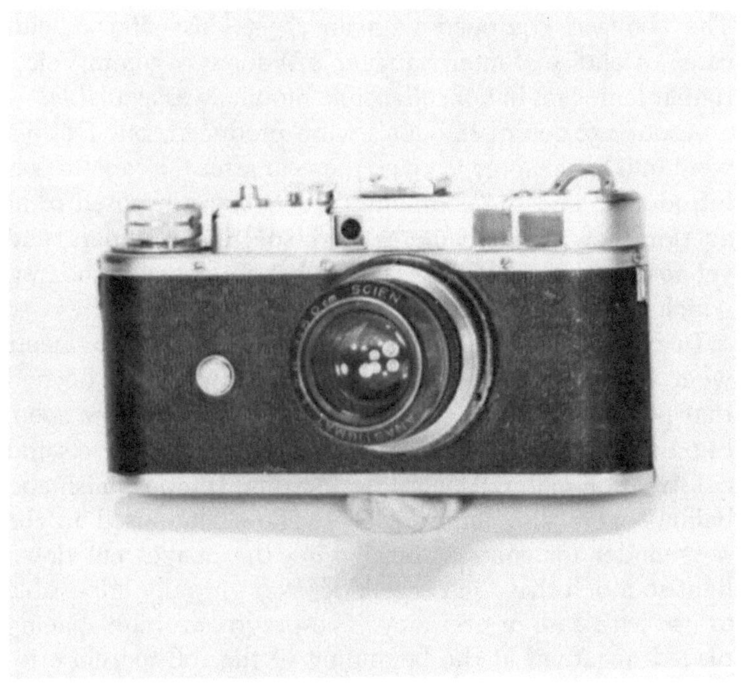

The Detrola 400 was as close to the Leica in appearance as any American camera made before World War II. The Detrola had several advanced features, designed as improvements on the Leica, such as a removable back and flash synchronization through a hot shoe. Unfortunately, the Detrola Corporation, which had successfully manufactured small radios, automotive accessories, and vest-pocket 127 cameras, was unable to obtain the necessary funds for the extensive advertising and precision workmanship which the camera needed. Within a year Detrola was bankrupt and sold at auction. The camera shown above was an engineering prototype and an extra accessory shoe was added to its top housing. The final camera had a Wollensak lens and lugs for a carrying strap.

The 400 was equipped with an *f*/3.5 lens ($69.50 with case). A choice of interchangeable Wollensak 50mm Velostigmat lenses set in noncollapsible mounts was available.

About 250,000 of the 400's were produced, but Detrola went into bankruptcy in 1941, the year after the camera was introduced, and the company's assets were disposed of at auction. Few of the cameras were sold to consumers, and yet no one seems to know exactly what happened to the rest, which must have been a rather large number.

In copying the Leica body design, Detrola's engineers went as far as they legally could and improvised beyond that point. As a result, the external design was very good, but mechanically the camera was very poor. The rivets and construction materials were far too thin. Interior finish and baffling were also inferior. The yellow filter used in the rangefinder for contrast in aligning the images cut down light so much that the rangefinder was virtually impossible to use with ease or accuracy. The progressive film spacing placed negatives at the beginning of the roll too close together and those at the end too far apart, making it impossible to mount slides automatically. In short, the Detrola 400 was characterized by poor workmanship in construction and assembly.

At the other end of the price scale were many inexpensively made, low-priced 35mm's, designed primarily to capture the "second-line" and punchboard markets.[12] One of these lines was the Falcon, introduced in the late thirties by the Utility Manufacturing Company, of New York. The Falcon line ranged from a simple $2.49 camera to the 35mm Super Action Candid, the top of its line, priced at $21.50. The Falcon 35mm had a molded-plastic body that closely resembled the Argus A in shape and larger features. It was equipped with an *f*/3.5 Wollensak Velostigmat lens in a helical focusing mount and a lever-set three-speed Deltax (single-action) shutter. The tubular viewfinder was made of metal, and beside it was an accessory clip. Distributed through drug, department, and sporting-goods stores and photo shops during the thirties, the Falcon line reappeared after the war under the Spartus marque, manufactured in Chicago by the Spartus Camera Corporation.

The Spartus 35mm body so closely resembled the Argus A series that the two could easily be mistaken for each other. The Spartus came in two models, one with an *f*/3.5 coated anastigmat lens in a micrometer focusing mount ($29.50) and the other with an *f*/7.7 achromat lens ($14.95). The former had a four-speed single-action shutter; the latter had a single-speed shutter but was synchronized for a plug-on flash unit ($4.50). Partly because it was also marketed primarily through mass-distribution outlets rather than photographic stores, the Spartus line lasted into the fifties, a good deal longer than its quality warranted.

[12] In states where it was legal (and usually where it was not), millions of individuals sold punches in boards they received in the mail. These 2 × 5-inch cards usually had about 20 holes, each identified with a girl's name. Inside the punch was a number which told how much the punch cost, generally 1 to 29 or 39 cents, and also whether one had won a lesser prize, such as a pen or a wallet. On the back of the card the buyer wrote his name next to the punch he had made, and the last person to punch also punched the grand-prize spot, which revealed the winning name. The grand prize was quite often a camera. The person selling the punches sent in all the money he had collected and received in return the lesser prizes plus two grand prizes, one of which he awarded, keeping the other for himself. Frequently the seller skillfully used a razor blade on the grand-prize spot to discover the winning name, which he sold to a friend or a relative—or himself—a practice which may help explain why punchboards are illegal in all states today.

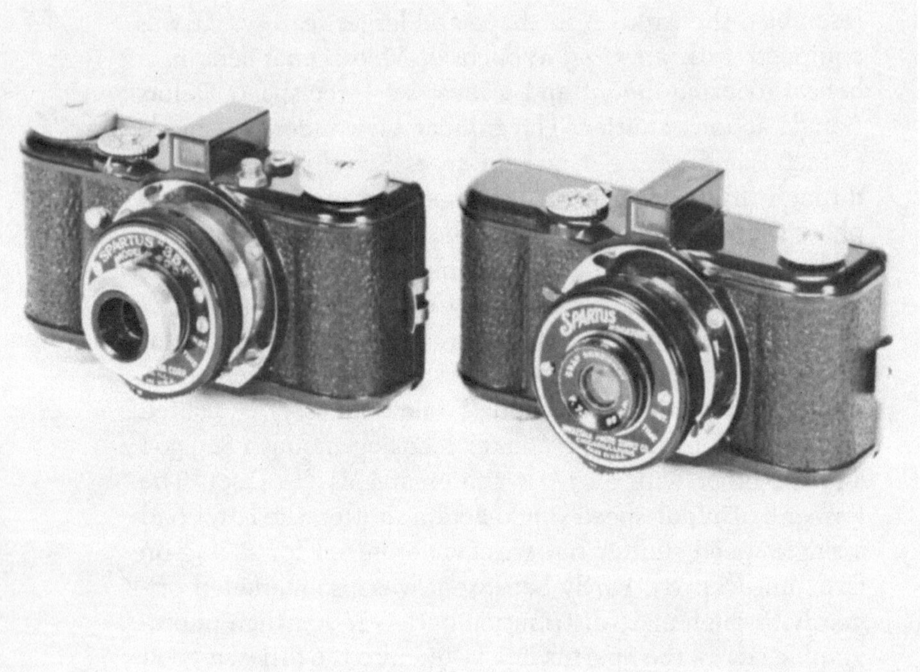

The camera above, under the names Spartus "35 F" (left), the Spartus Miniature (right), and others, was produced by a company that went on to find its metier in clock manufacture. It appears that the original dies were bought from the Utility Supply Company, of New York, which made the prewar Falcon line of box cameras. After the war Spartus made a full line of inexpensive cameras for the punchboard and carnival trades. The most famous of the line was not a 35mm but the Press Flash, a 620 box camera with a built-in flash reflector. The 35mm was made in several versions, many carrying private-label names. Most of the cameras were mechanically poor, and some had unsound engineering developments such as aluminum shafts through aluminum bushings.

While all but die-hard photo fans have forgotten the Winpro 35mm, in its time it was an adventurous undertaking for Claude Wright, J. A. Henne, Jim Wilmot, and Jim Morrisey. The story opened in 1946, when Chester Crumrine, an independent camera designer and one-time engineer for Kodak, was asked by William J. Brown, of the Kryptar Corporation, of Rochester, New York, to design a small 35mm box camera that he could sell with his film line. (In the years to come, Brown would suffer two disastrous fires which would end Kryptar, and from its ashes would arise the Dynacolor Corporation—but that jumps ahead of our story.) Brown soon found that the money he had invested in the design was more important than the camera, and he sold both the design and the concept to Jim Wilmot, a Rochester, New York, real-estate broker. Wilmot in turn interested other partners. Soon, in a low-rent district of Webster, New York (near "Camera City"), Webster Industries went into the 35mm-camera business.

Henne, a camera engineer, found that making the camera was more complicated than merely following the design the partners had purchased from Brown. In the end about all that remained of the original design was the general appearance of the camera and the unusual under-the-body frame counter. Henne designed and patented a unique flywheel-governed shutter which "wound up" before the escapement made the exposure. No matter how the shutter was activated, the exposure was consistent, an uncommon characteristic in simple cameras. The two-element fixed-focus lens was rated at *f*/7, and the 40mm focal length gave the pictures a slight wide-angle effect.

At first Webster bought its lenses from a two-man optical firm across the street from the company, but when that firm went bankrupt, Webster purchased machinery and began producing its own lenses. General Electric made the molds for the Tenite plastic body. This thermal plastic made unique demands on the mold designers, but they managed to produce a body that was satisfactorily stable.

The Winpro 35mm was a good camera for the price and time, but, like many before them, its makers envisioned another Argus boom. Some of the potential was there: simple cameras using 35mm film at a price below the magic $15 limit were in short supply. Arrangements were made for the camera to be marketed by national photographic distributors through Caradell Products Company, of New York City. As a publicity stunt the plastic camera was dropped from a skyscraper window to the sidewalk to demonstrate its unbreakability. Available news reports do not record any disasters.

The first Winpro 35mm cameras were without flash synchronization, but they sold reasonably well, and when it became evident that flash was a desirable selling feature, flash synchronization was added, though a nonflash model apparently continued to be manufactured as a promotional item. Between 1947 and 1949, Webster made and sold more than 150,000 of its inexpensive 35mm cameras. But late in 1949 it was apparent that the bloom was off the Winpro, and the company's remaining partners, Morrisey and Wilmot, announced that they would move operations to other company-owned property in Horseheads, New York.

Herb Pfeiffer and Morris Cassorla, partners in Monroe Warehouse Company, of Rochester, were convinced that a camera like the Winpro could make real money if it was properly marketed. They bought the dies in 1953, farmed out production, renamed the firm Monroe Research (a few months later the company became the Zenith Film Corporation), and began peddling cameras. The questions posed by retailers were formidable. Was the camera better than a Spartus that looked like a more expensive camera? Would it outsell a 620 Kodak Brownie Hawkeye, especially now

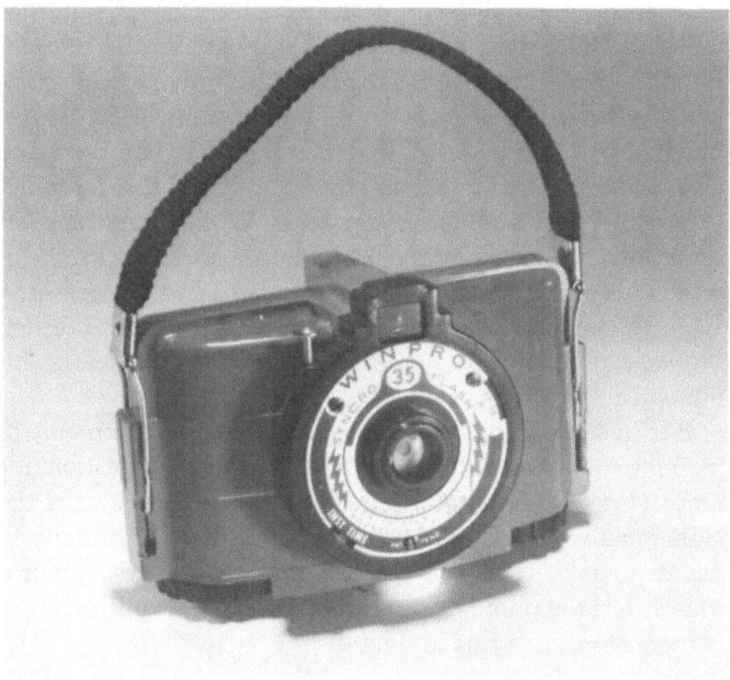

Manufactured of Tenite and incorporating an f/7 lens, the Winpro 35 Syncro Flash enjoyed a substantial success with the snapshot trade after World War II. Its low price and high quality (for an inexpensive camera) made the Winpro a favorite, especially after the introduction of low-cost color-slide film, but, like most independent ventures in the photographic trade, time and inadequate financing finally brought its manufacture to an end.

that Kodacolor film was coming into its own? The answers Pfeiffer and Cassorla gave were not too persuasive.

The partners reached the rather surprising conclusion that, while people would buy the inexpensive camera, they could not afford the color film. What was needed was an inexpensive color-slide film to go with the camera, and that is where Bill Brown and Dynacolor reentered the picture. By that time Dynacolor Corporation, Brown's new company, was well on its way to becoming the world's largest independent processor of Kodachrome film. The company also made Dynacolor, a color-slide film based to a large degree on expired Kodachrome patents. While Dynacolor left something to be desired in comparison with Kodachrome, it was inexpensive, and Brown was happy to find a distributor. It particularly pleased him that the camera whose design he had commissioned should be the vehicle for selling his film.

But luck was not on Zenith's side. The old problems of inadequate financial structure and products of questionable merit (the new firm had farmed out the manufacture of the camera and was plagued with problems of quality control) began to take their toll. In less than two years Winpros quietly left the photographic scene as closeouts in Rochester camera stores at $7.95 and less.

Other inexpensive postwar 35mm miniatures made brief appearances. Among them was the Classic "35," a small aluminum half-frame camera made by the Craftsmen's Guild, of Hollywood. But increasing competition from Japanese and German imports made their existence precarious. By the close of the Korean War in 1953, the Japanese had replaced imitation with creativity, and they soon

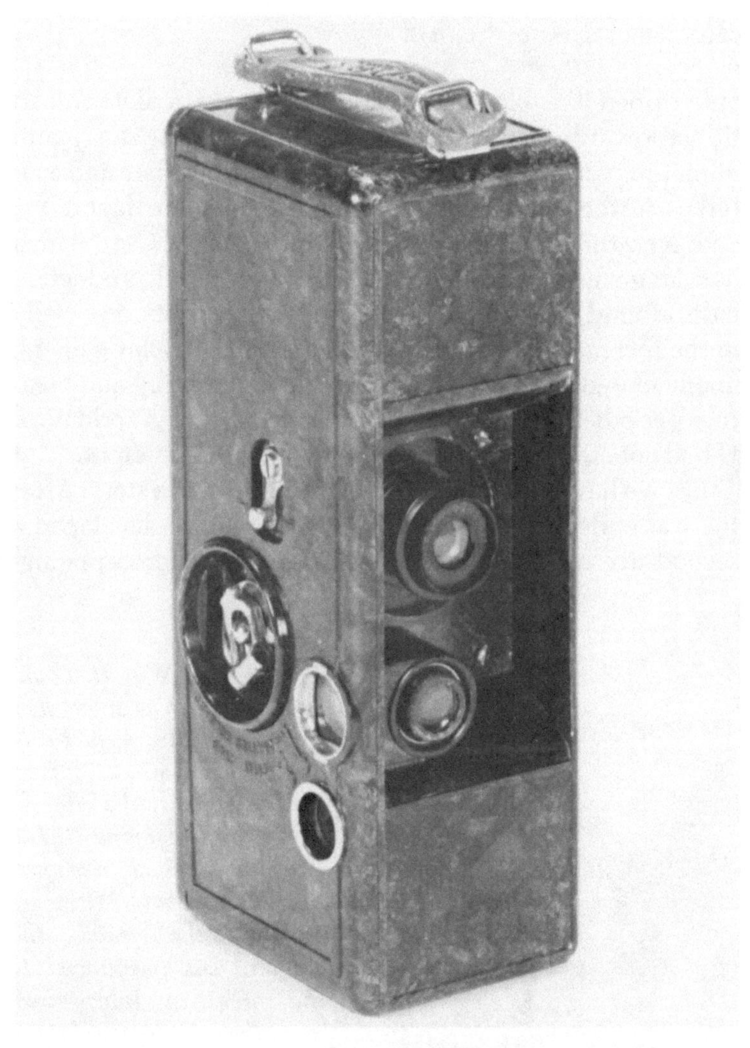

Besides its Bakelite construction, the QRS is both an odd and a mystery camera. Little is known of its background, development, and demise. The company was more famous for making player-piano rolls. It later became QRS-DeVry and went on to teach electronics by mail.

outstripped the older German industry in a determined thrust toward dominance. The age of the American 35mm miniature was nearing its close. The cost of labor and materials in the United States could not compare favorably, and even time-honored designs like the Argus C-3 (which had long since amortized its design and initial production costs) found the going rough. New developments, especially in the increasingly popular 35mm single-lens reflex camera, finally closed the door on one of the most exciting and creative periods in the history of photography. If World War II had not erupted, the story might have been different.

But we have not quite reached the close of the story. After the war ended, the American 35mm miniature developed a second eye on the world, and in the fifties stereo photog-

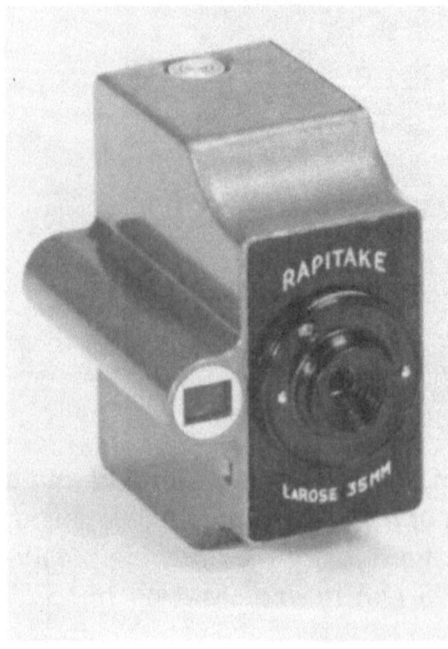

After World War II would anyone design a camera that would take the Agfa Karat cartridge? German manufacturers did, and so did an American, Raymond R. LaRose, who had designed the subminiature Micro 16. Known as the Rapitake, the camera was introduced to the press in July, 1948, in Culver City, California. Neatly designed and compact, the Rapitake featured a fixed-focus lens with one shutter speed. It promptly disappeared.

raphy enjoyed a healthy though brief vogue, resulting in the design and production of several interesting 35mm stereo cameras. Read on.

17 Twin Windows on the World

The stereoscope preceded photography. As early as 1550, Galen investigated the idea of using two flat images to produce one image with true dimensional perspective. Several months before Louis Daguerre and Fox Talbot made public their announcements in 1839 (the year which photographic historians use as the beginning of the art), Charles Wheatstone addressed the Royal Society on the subject of the stereoscope. Since that time application of stereoscopic principles to photography has been an on-and-off affair, with stereo booms appearing in cycles, usually after major conflicts, such as the Civil War, the Spanish-American War, and both world wars.

Stereo photography is based upon the fact that when we look at an object each eye sees it from a slightly different angle. The brain superimposes these two pictures, making them one and allowing us to judge depth. The usual camera has but one lens and makes a single picture, which reduces depth to a single dimension. But when two lenses are positioned side by side with a separation equal to that of human eyes (about 2½ inches), two pictures result which closely approximate what the human eye sees. When one concentrates as he views the finished pictures side by side from a distance of about 14 to 16 inches, the two gradually merge into one, and perspective appears.

The parlor of the turn-of-the-century American home was graced with a wooden stereoscope (a Brewster or a Holmes), along with boxes of cards on each of which were mounted two almost identical pictures. It was a pleasant spectator sport for winter afternoons in the early 1900's. But as a participant activity stereo photography has had its ups and downs and has never held the public's fancy for long. The lack of a complete stereo system, the inconvenience in projection, the high costs involved, and the difficulties in mounting and viewing the pictures have all played roles in the fickleness of camera fans.

Our interest in the stereo story begins in 1941,[1] when the David White Instrument Company, of Milwaukee, Wisconsin, a long-term manufacturer of precision surveying equipment, began developing a concept which would culminate in a complete 35mm stereo system.[2] Designed by Seton Rochwhite, a lighting engineer, the Stereo Realist and its accessories appeared late in 1947, setting off a stereo craze which lasted nearly a decade and inspired a host of

[1] At least one American 35mm stereo camera, the Depthro Stereo, was announced before World War II. This camera took 10 to 12 exposures on an 18-exposure roll, giving a picture size 15/16 × 13/16 inches (2 stereo pictures on each 5 inches of film). It had matched 2-inch *f*/4.5 Wollensak Velostigmat lenses in Alphax shutters (1/25 to 1/200), fixed focus, and a die-cast aluminum body. The Depthro was said to have an "automatic film winding mechanism and built-in flash synchronization," and the makers claimed that the "pictures need no transposing for projection." It was offered by Stereovision, of Claude, Texas, at a price of $95.00. The camera was pictured in the *Popular Photography Directory* for 1941, along with a $79.50 projector.

[2] The David White Company changed ownership in 1959, the name became Realist, Inc., and the company moved to Menomonee Falls, Wisconsin. Wisconsin might well be called the "Stereo State"; at least three cameras, the Realist, the Contura, and the Kindar were manufactured there.

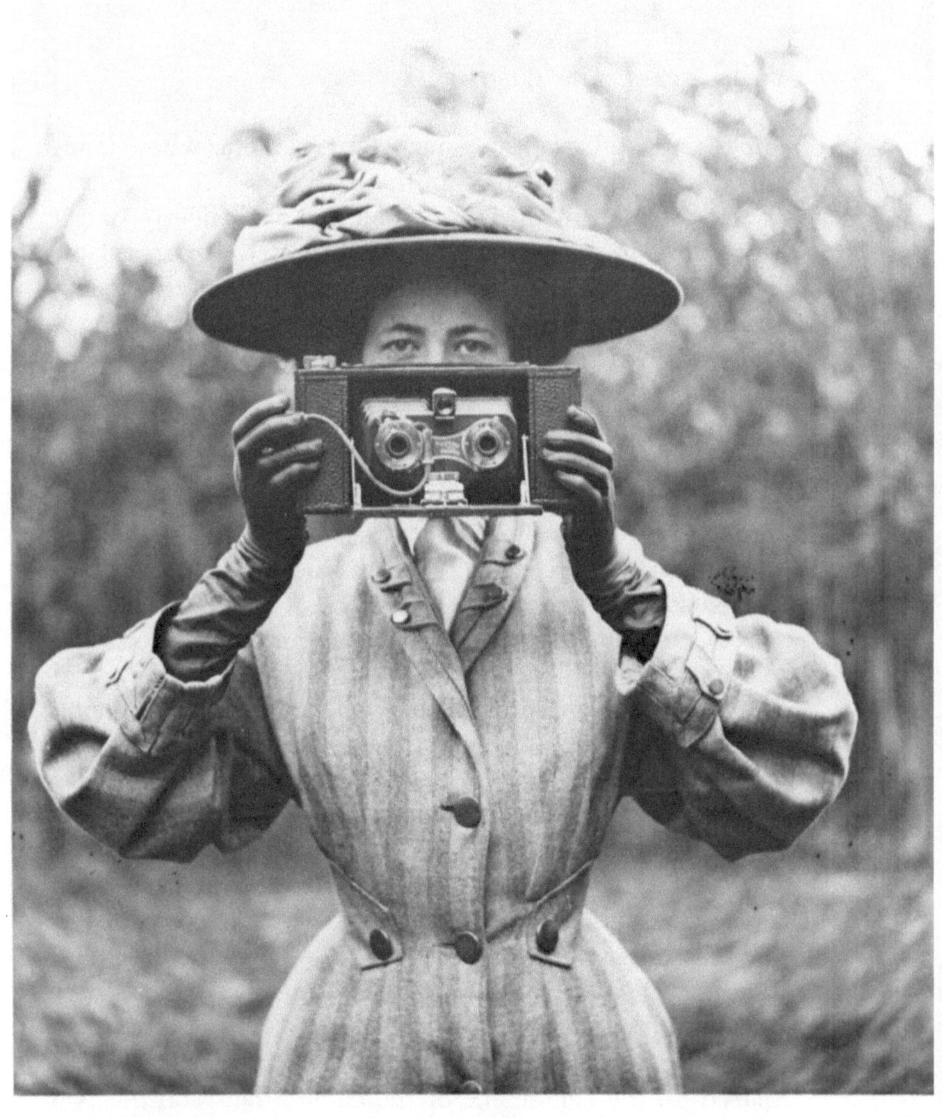

In 1916 stereo photography was not exactly "candid," as this charmer using a No. 2 Kodak Brownie Stereo illustrates. The pictures she took (on roll film) could be viewed in standard parlor viewers found in most homes at the time.

American and European 35mm stereo cameras whose manufacturers hoped to share in the market.[3]

Even so, the Realist never had a real challenger (American or foreign) with which to lock horns in the marketplace. The camera was in a class by itself, and shrewd advertising, coupled with a top-quality product, made its market position invincible. According to competitive surveys, testimonials by famed personalities were very effective and an unusually well managed asset, but the camera's greatest boosts came free. On one of its 1952 covers *Life* featured a photograph of General Dwight D. Eisenhower, and Ike just happened to have a Realist hanging around his neck. Bitten hard by the stereo bug, Harold Lloyd, a silent-screen comedian, gathered a large collection of stereo slides as he traveled around the country promoting the Shriners. His slide shows were well-attended expositions of the art of stereoscopic photography. They won for Lloyd the unofficial title "Mr. Stereo" and gave the Realist a boost unobtainable through conventional advertising. Best of all for the Realist, Lloyd did it without charge.

The Realist camera was unusual in appearance and had a left-hand shutter release. Rugged and solidly built of metal covered with black leather and chrome trim, it was one of the few stereos of its era which possessed an aura of quality. Ilex furnished both the 35mm dual matched *f*/3.5 lenses and the interconnected flash-synchronized leaf shutters (1 to 1/150, *T* and *B*). To avoid parallax, the viewfinder was placed on the center line of the camera between the two

[3] The advent of 35mm color film made this particular stereo boom possible. While the View-Master came along too late to reap the full potential of the market, inexpensive mirror adapters like the Stereo-Tach were made available for nonstereo 35mm and reflex cameras. Leitz and Contax brought out their own highly corrected adapters.

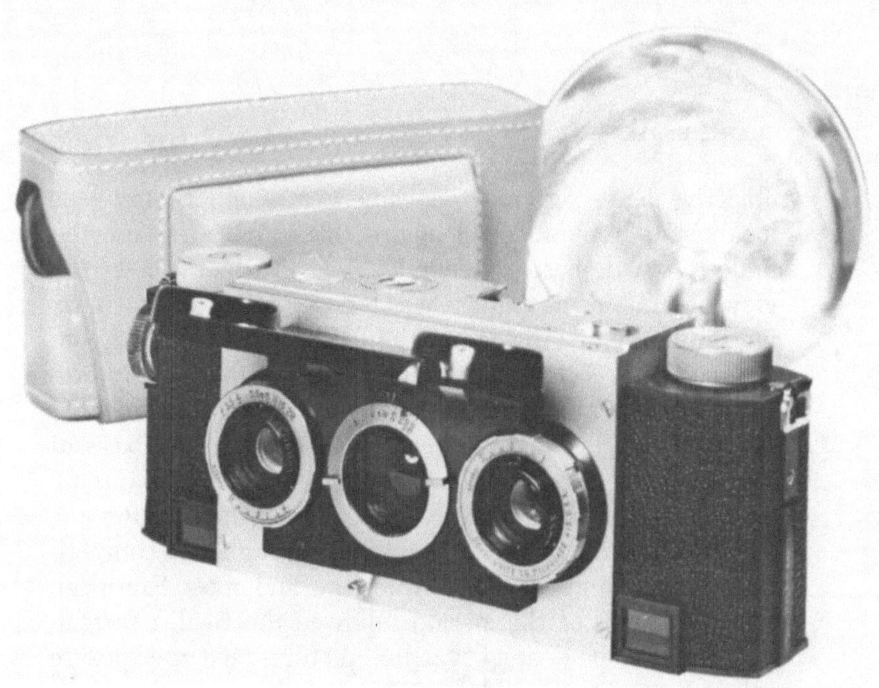

The postwar stereo craze was started by the Stereo Realist. It was unusual in design: it was focused by moving the film plane rather than the lenses, and the rangefinder was on the bottom of the camera. It also had an unusual-size stereo format. In spite of, or perhaps because of, its differences, it set new standards for stereo which manufacturers all over the world soon adopted. Only the French stayed with the larger format, just as they had retained 9.5mm movie film after 8mm became standard. The model 1041 shown above had matched f/3.5 Ilex lenses in shutters with speeds of 1 to 1/150 second. A special f/2.8 model with Kodak Ektar lenses had one production run of 2,000 cameras. A special run was later made for Olson Camera Company with lenses remaining from the initial production. The drop in stereo popularity prevented a planned follow-through for the general public. Some of the f/2.8 models had brown-cowhide coverings, but very little change was made in the camera itself beyond altering the depth-of-field scale and adding a double-exposure prevention with override at an early production stage.

objective lenses, giving the camera a peculiar three-lens look as the user braced it against his forehead to use the finder. A wide-base rangefinder was incorporated in the viewfinder, and the camera was focused internally by moving the film plane instead of the lens.[4] The Realist's 70mm *ocular distance* (lens separation) was not that of any previously accepted standard (usually 63mm for a 35mm camera) but appears to have been established on the basis of film economy and a reasonable compromise in separation. The Realist was far and away the leader by the time Revere, Kodak, Graflex, Universal, and others brought out competitive models, and all other American and most European stereo cameras of this period followed the Realist format, producing 16 pairs of 23 × 24mm pictures on a 20-exposure roll of Kodachrome and 29 pairs on a 36-exposure roll.[5]

Except that double-exposure prevention was added and the depth-of-field scale was simplified, the original Realist design was altered only slightly over the years. Another model, with *f*/2.8 Kodak Ektar lenses in shutters ranging to 1/200, was produced, but the Realist fan had to look closely to tell the difference. Some special offerings were also manufactured, including a model covered with brown California saddle leather. In a special deal the cameras were offered to photographic dealers and clerks with their names engraved on top. This model was bootlegged throughout

[4] The rangefinder had a design defect similar to that of the Argus C-3 (see Chapter 9).

[5] Exceptions were the French Verascope, manufactured by Jules Richard and distributed in the United States by Busch, and the Keys Stereo (manufactured by the James V. Keys Company—later Keys Stereo Products), which used 828 film. Toward the end of the stereo boom, Lionel (the toy-train manufacturer) and others brought out their own odd-sized models. All quickly failed.

New York; Realist sold the camera, case, and viewer in a set for $100, and dealers promptly sold them to customers willing to buy a new camera with someone else's name on it at a substantial savings. Since the package cost was considerably below the usual price, dealers were able to make a nice profit, ordering one for each clerk in their store (and

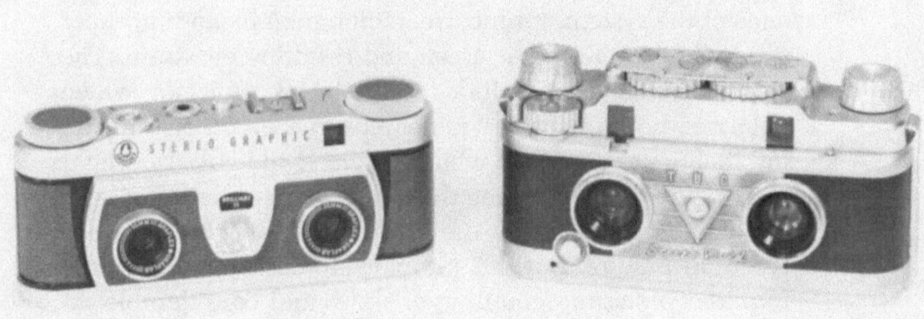

Besides the Stereo Realist, two other American stereo cameras were produced at the height of the boom and had a reasonably good reception. The Stereo Graphic (left) was an elongated Graphic 35, and the two shared many interchangeable parts. A basic camera, it has the simplicity of line that designers particularly liked. The TDC Stereo Vivid (right) which looked like a foreign camera (and had a companion model that was foreign-made), sported a rangefinder and an unusual front exposure button (note the similarity to the View-Master Personal Stereo). The TDC sold fairly well, and the company was later purchased by Bell & Howell.

probably several for nonexistent clerks) and selling them over the counter.[6]

Not an inexpensive camera ($149.50 with the *f*/3.5 lens, $215.00 with the *f*/2.8 lens), the Realist was only the first step in a complete system. The David White Company also made slide viewers with matched lenses and a compatible mounting system which assured the user of proper slide alignment.[7] This detail was very important for the acceptance of the system, for incorrect alignment meant improper superimposition on the screen and resulting eyestrain. The Realist system also included a dual lens projector which polarized both images at 90 degrees to each other, projecting them onto a surface which reflected the polarized light to the eye. Those watching the screen wore polarized glasses so that the left eye saw the left picture and the right eye the right picture. But this ultimate stereoscopic experience required precision mounting of slides and considerable expenditure of time and money. It never really became popular except with the purist.[8]

[6] After the Realist was discontinued, Olden Camera in New York negotiated a deal with the company, which made a last production run of approximately 80,000 new Realists using *f*/2.8 lenses. The same arrangement was made with the manufacturers of the German Iloca II Stereo, which Penn Camera of New York distributed for some time after production of the camera had been halted. These lenses were unmatched—that is, not bench-calibrated for identical focus. Unmatched lenses gave a magnified image which made projection impossible and viewing difficult. Cameras such as this contributed to the decline of interest in stereo photography.

[7] Realist stereo viewers were apparently copied by just about everyone in the business, who changed the external appearance sufficiently to make them look different. Internally the Wollensak, Revere, TDC, and others were identical to the Realist viewer, which was priced at $24.50 (battery model) and $29.95 (electric model).

[8] Stereo also found much favor for commercial use. Salesmen carried hand viewers and slides to display their wares, and wedding photographers had a much-desired and very profitable service to offer in addition to

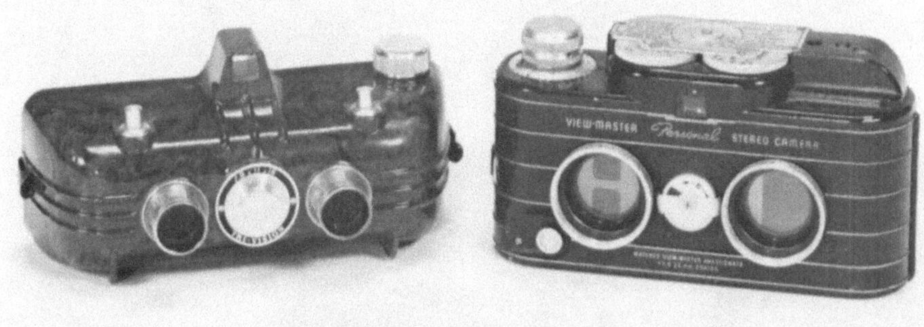

The Haneel Tri-Vision (left) was a camera that seems to have been reincarnated at frequent intervals despite its extremely poor engineering and fabrication. Also known as the Keys Stereo, as well as by other names, it seems to have always been made of an unstable plastic that warped. The frames around the film plane were uneven, and the two openings were not even the same size. Two exposure plungers allowed for stereo or single pictures. The first View-Master Personal Stereo camera (right) was an attempt to make the popular View-Master reel system available to the amateur photographer. Although the camera was well built and extremely rugged, it had an over-all unattractive appearance and was not successful. The company brought out a less complicated German-made model with a simplified film transport which is still available. The manufacturer, Sawyer's, Inc. (now a part of the GAF Corporation), was the first independent processor of Kodachrome film.

their usual coverage. The Photographic Society of America recognized the importance of stereo by establishing a stereo division in 1951, a move not made lightly by that rather august organization.

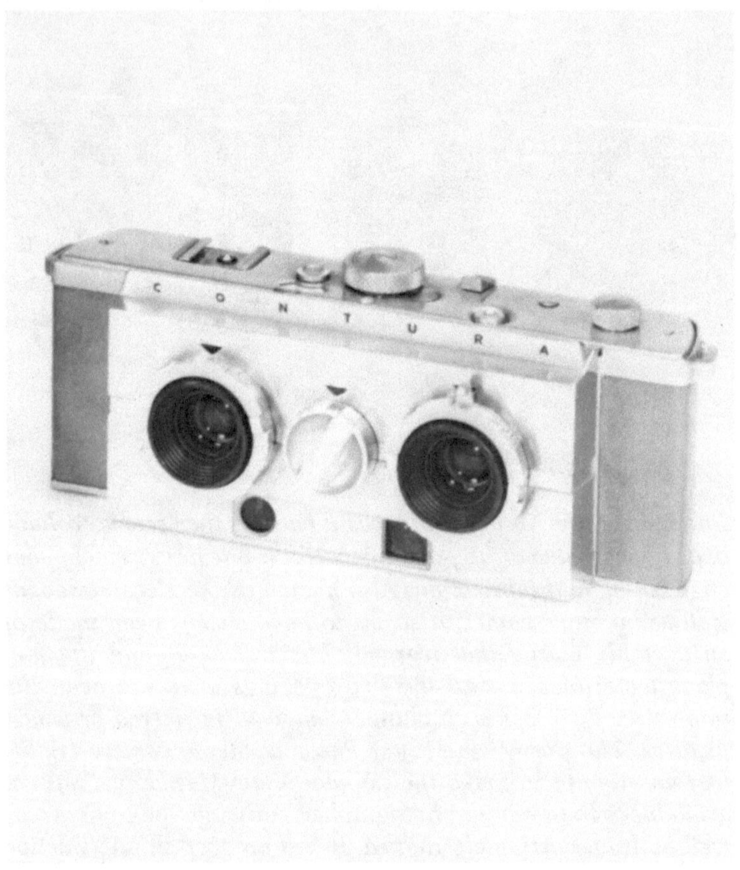

Perhaps the most beautifully designed American stereo camera is the almost-unknown Contura, made by the Stereo Corporation, of Milwaukee. It resembled the Stereo Realist, including the bottom rangefinder. The lenses and the shutters appear to be of Ilex manufacture. The shape, the center winding knob, and saddle-leather trim make it an unusually graceful camera in appearance. Note the left-hand shutter release button above the R.

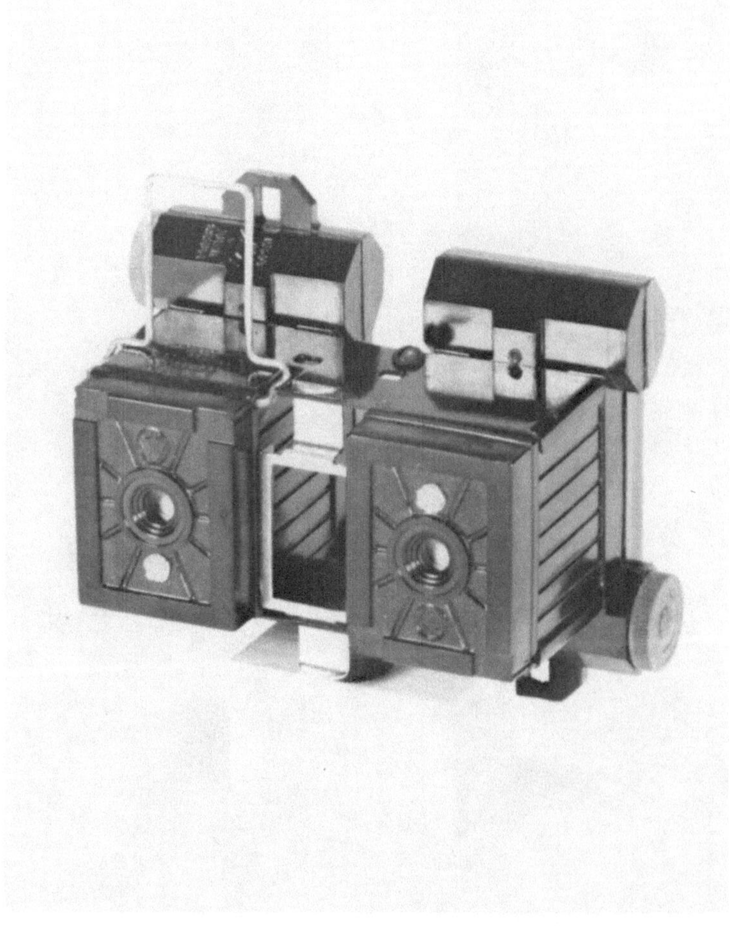

Never underestimate the originality of the American entrepreneur. Take two 39-cent cameras, turn one upside down, add a spacer that also trips the shutters simultaneously, and presto—a stereo camera! This little adaptation of the original UniveX A was the product of an imaginative Californian.

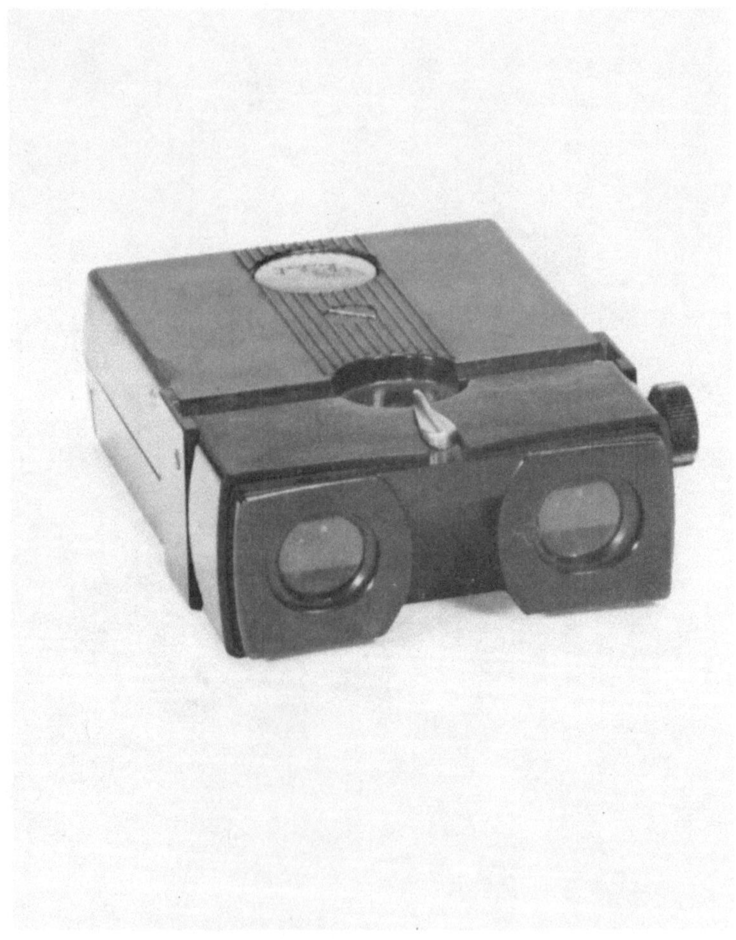

The stereo viewer with which all others are compared is the Stereo Realist. The company also made less expensive viewers, and it is believed that they made the Revere viewers as well, for inner construction appears to be identical. This viewer provides for interocular as well as focus correction.

The stereo boom hit its peak in the early fifties, when several other American cameras appeared. Most, like the Graphic and the Universal Steré-All, were limited in range and quality. The Revere, introduced in 1952 at $179.50 (and also marketed under the Wollensak name with *f*/2.8 lenses), matched the original Realist in specifications but failed to equal the quality, feel, and operation of Rochwhite's design. One of the two stereo cameras marketed by the Three Dimension Corporation (TDC Colorist and StereoVivid, later Bell & Howell) was manufactured in Germany and thus falls outside the scope of this book, but neither was a contender for any particular honors. This left only one other American stereo camera of any consequence, the Kodak Stereo.[9]

Designed by Douglas Harvey, the Kodak Stereo almost failed to see the light of day. Believing that the stereo craze had about run its course, Dr. Albert K. Chapman, then president of Kodak, initially said no.[10] Dr. Chapman took the

[9] The View-Master Stereo, which found brief favor in two models (one American, one German) was marketed to accommodate View-Master users who wanted to produce their own reels. All necessary equipment for home mounting was made available, including blank reels holding the 12 × 13mm slides, but it was a task of major proportions, even more so than mounting Realist slides. The cameras took 37 or 69 pairs on a roll of 35mm film, and while Sawyer offered a mounting service, the cost was all but prohibitive except for the dedicated. A well-thought-out system, this one definitely died from the cost and time involved in it. Graflex made the Graphic Stereo, using many parts from its Graphic 35, and Universal had its own Steré-All, but neither came in time to benefit.

[10] With the advent of Kodachrome, Kodak resurrected the design for a stereo camera which had been proposed sometime before, refined the concept, into the first American 35mm stereo camera, a full-frame, all-metal stereo. Production was about to begin in 1940, but was halted because of the war. America's defense requirements and later entry into the war led to the abandonment of the program. This fact was not disclosed until after the war, when Kodak dealer publications revealed that the Stereo 35 of 1940 preceded the new Kodak model.

position that stereo photography enjoyed a boom every generation, lasting six to eight years and then fading. Presented with a much-discussed and apparently sound case for the camera, however, Chapman finally relented, giving Harvey a free hand.

Introduced at the close of 1952 at $84.50, the Kodak Stereo bore only a faint resemblance to the Realist. It had a Kodak Flash 200 Dual shutter (1/25 to 1/200 and *B*), and three-element 35mm *f*/3.5 Anaston lenses separated by a viewfinder placed on the center line of the camera. The viewfinder contained a spirit level, a useful accessory which helped the photographer make certain he was holding the camera correctly; tilting the camera made it necessary for the pictures to be viewed in the same manner, which disoriented the viewer.

All settings were visible from the top of the camera. The body was finished with a neat tan leatherette covering with metal top, bottom, and back plates. Of all the stereo cameras introduced during this decade, only the Realist and the Kodak had the feel and appearance of quality. Although the Kodak Stereo enjoyed good sales for a short time, Chapman's intuition proved to be correct, the stereo boom was suddenly over, and the Kodak Stereo languished.

What killed stereo? Many answers have been offered, but most agree that the time, the cost, and the difficulties involved in mounting and projecting stereo slides overwhelmed the users. Although processors, including Kodak, mounted the slides (and still do) in paper mounts for hand viewers, the mounts were not practical for projection purposes owing to the frame separations. Projection requires exact mounting according to scene focus—close-up, medium, or infinity. Each of these positions requires a different sepa-

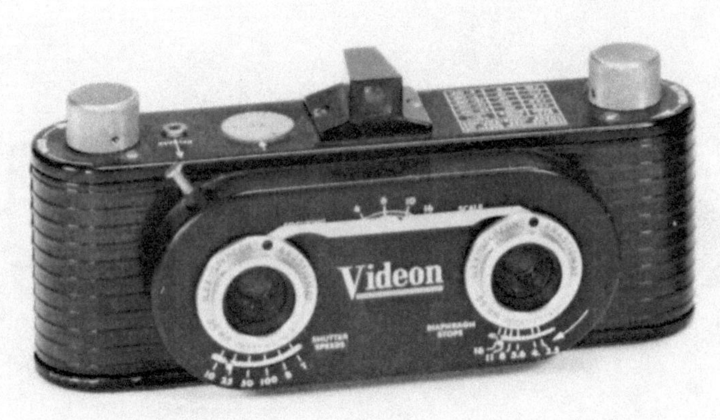

If you have seen any flying saucers lately, get in touch with the air force. They may supply you with a Videon. Reportedly designed by a group of disgruntled former Realist employees as a poor man's Stereo Realist, it had the same lens-and-shutter system but with front focusing and no rangefinder. During the flying-saucer craze, the air force equipped the camera with a grid that was attached to one lens, enabling a photogrammetrical evaluation of sightings. Apparently no reports of results are available.

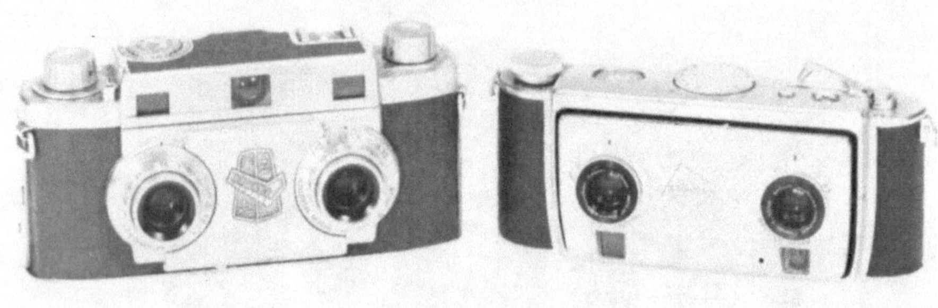

Two cameras that rode the wave of the postwar stereo boom—one a success and one a near miss. The Revere (left) and its sister Wollensak (the latter with f/2.8 lenses) moved into the market early and enjoyed relatively good sales. Well designed and sturdy, they were traditional cameras without the innovations or the popularity of the Stereo Realist. The Kin-Dar (right) was a sleek, compact stereo camera that received a boost from a story about it in Life. *Very much like a streamlined Realist, it entered the market late and, along with the Kodak Stereo, failed to achieve the success it deserved.*

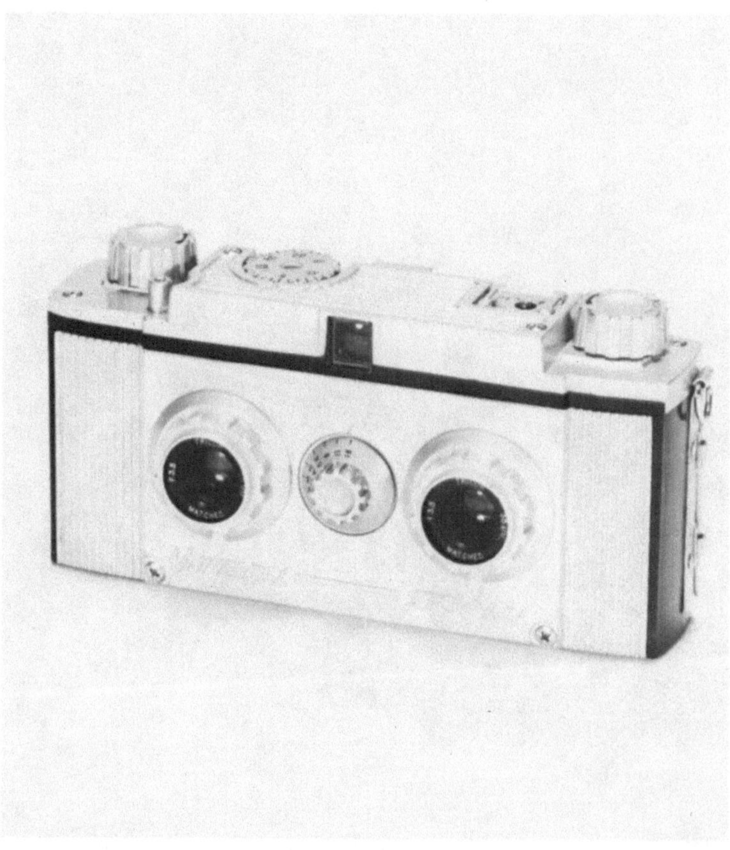

Universal's contribution to stereo photography, the Steré-All, was its last camera and a mockery of previous designs. Big, bulky, and all-plastic, it was at least completely original and totally different from other Universal models. The Steré-All looked inexpensive but apparently worked well. In 1964 one of the authors met a tourist in Europe who was making his third trip around the world snapping excellent stereo pictures with his Steré-All.

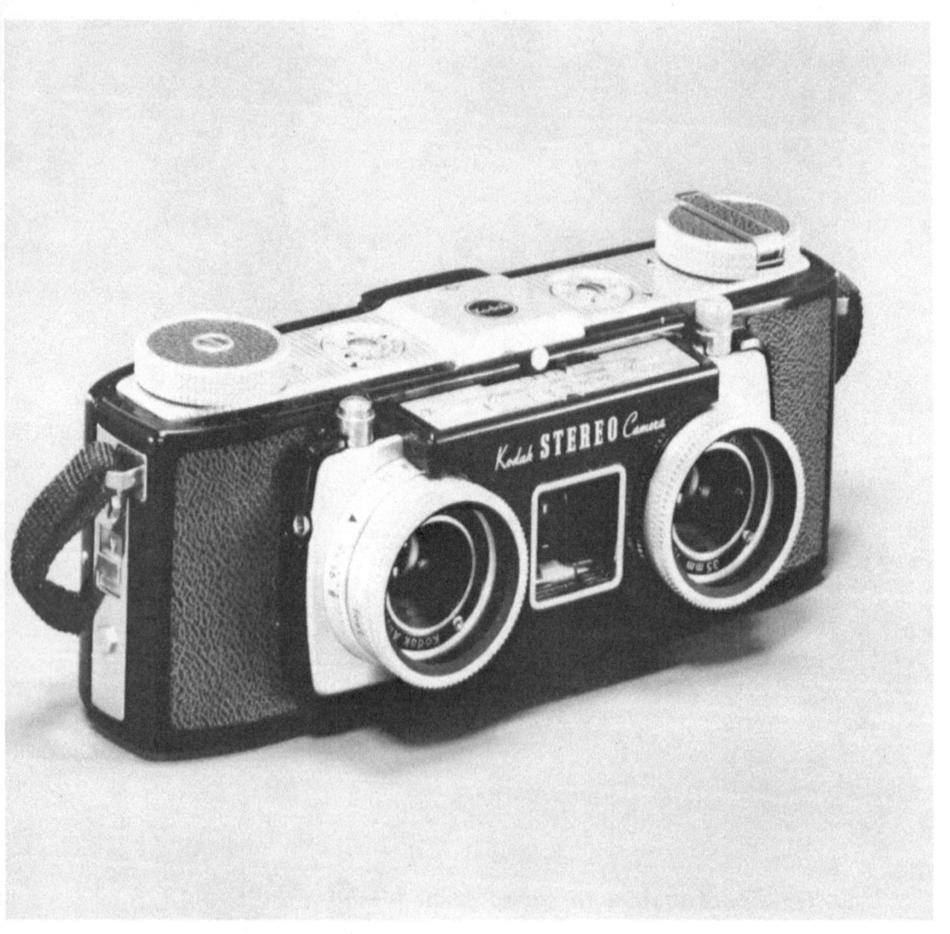

The well-built Kodak Stereo Camera had simple yet clever features, such as a built-in level, but it was marketed near the end of the postwar stereo boom, and sales were disappointing. As the stereo miniboom of the 1970's began, a good used Kodak Stereo was selling for a price almost as high as the cost when it was new ($84.50).

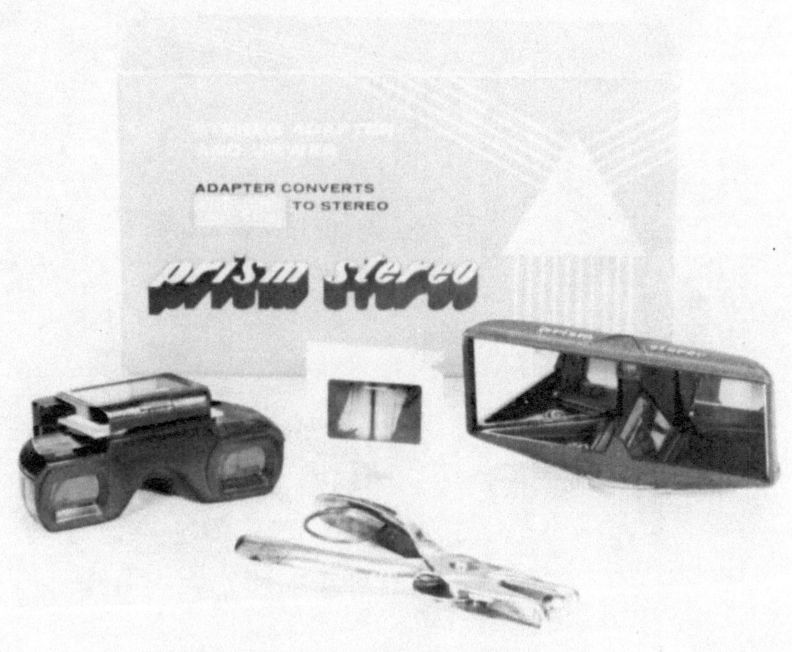

Many adapters were designed to convert a regular 35mm camera to stereo. Most of them produced double images on a standard double frame, although a number of parallel-arm platforms were available to shift the camera to make two-frame stereo pictures. One of the more interesting adapters was the relatively obscure Prism Stereo, made by a New Jersey firm. The Prism Stereo appears to have been well designed and manufactured. It came complete with a small viewer and a punch to help align the slide in the viewer. The user had to hold the camera with the lens upright to take the picture and sighted through a tiny viewfinder built into the adapter.

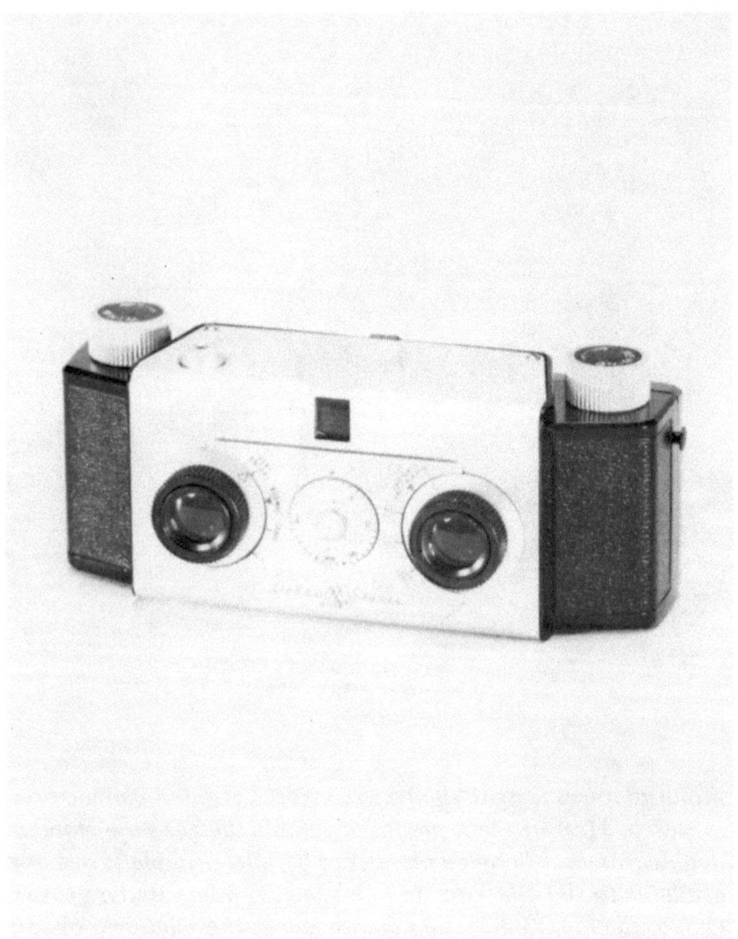

Door-to-door camera selling is rare, but the Delta Stereo was sometimes sold that way. A neat, basic camera, it has an unusual feature in that the rear eyepiece is also the lock on the back. Another unique design idea allows the lens focus to be adjusted for close-up flash pictures.

ration between the two images for correct superimposition on the screen by the projector. Thus slides mounted and returned by the processor had to be remounted by hand for projection. Projectors were very expensive (up to $700) and were bulky and clumsy in operation. The stretched-rubber screen with its aluminized surface had to be absolutely taut and perpendicular to the projector's lenses. With some systems viewers had to wear polarized glasses during projection. They were expensive and inconvenient and were forever being lost or broken, and there never seemed to be enough to go around.

All these are valid reasons for the demise of the three-dimensional craze (and did indeed kill stereo photography for one of the authors—Lahue). But another interesting and seldom noted trend in the postwar years has been the gradual displacement of color transparencies in public favor by color negative film and the resulting prints. At this writing color slides appear to account for only one-third of all amateur color pictures, which may be explained by the fact that women now take more pictures than men do.

Whatever the reason or reasons, stereo photography is once more quiescent. The Realist was discontinued some years ago after its competition disappeared, but the manufacturer continues to service the cameras and quietly but confidently expressed to the authors the expectation of a revival of stereo photography "in the near future." Such optimism should not go unrewarded, although we are willing to bet that should another stereo craze sweep the country it will use the instant-loading format and not 35mm—which brings us to the finale of our story.

18 Rebirth

As the fifties drew to a close, the American 35mm miniature camera was reeling under the combined onslaught of foreign imports and single-lens reflex cameras. The American camera industry had to accept the fact that it could not manufacture precision cameras in this country at a price low enough to compete with Japanese- and German-made cameras. Many manufacturers fell by the wayside in an attempt to do just that, while others tried to survive by marketing foreign lines under their own names. But the *coup de grâce* was delivered on February 28, 1963, with the worldwide introduction of the Kodak Instamatic system.

Kodak's interest in a single, foolproof, inexpensive photographic system that would increase film sales dated back to the early days of Dr. Kenneth Mees, whose concept of the Aunt Molly camera predated the development of the 35mm. Work on the integration of components in a unitary package began before World War II but was spasmodic and isolated. After the war Kodak engineers, like those of every other photographic manufacturer, were busy with the problems of reconversion and later the Korean conflict, but in the late fifties designers turned back to the system concept in earnest.

Engineering design of the Kodak Instamatic, under the name Project 13, was completed in 1961, and tooling started the same year. Production began in 1962. Project 13 was probably the company's best-kept secret of the century; comparatively few even within Kodak were aware of its existence. When the Kodak Instamatic line was presented to its salesmen, they were as flabbergasted as its competitors.

The Kodak Instamatic did not represent a genuine scientific breakthrough in photography, but it was a masterful example of engineering ingenuity and packaging.[1] Its most

[1] Even the least expensive of the Kodak Instamatic cameras utilize an *f*/3.5 plastic lens, which can be stopped down to *f*/11 or *f*/16 for better definition—a practice which the Ansco Memo had adopted in the thirties.

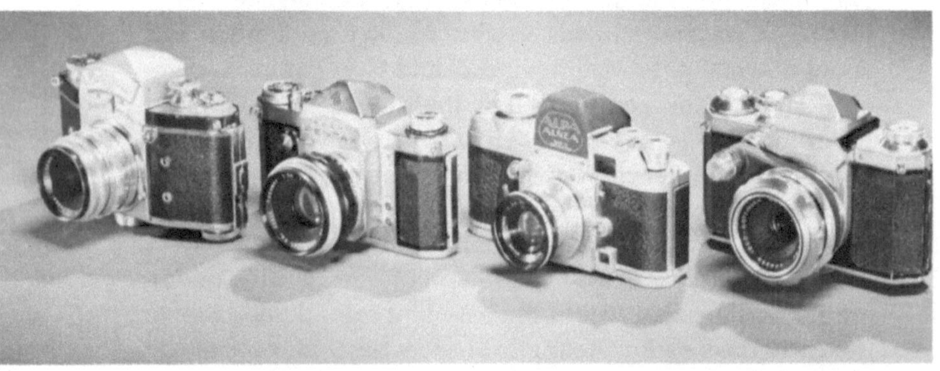

Four reasons for the decline of quality American cameras are these products produced by foreign manufacturers in the 1950's. Left to right: The dated but popular Exakta, the Asahi Pentax (later to become the very popular Honeywell Pentax), the limited-production Alpa Alnea (which had started life as the Bolsey Reflex), and the German Edixa Reflex. They were representatives of the new wave of interest in the single-lens-reflex camera.

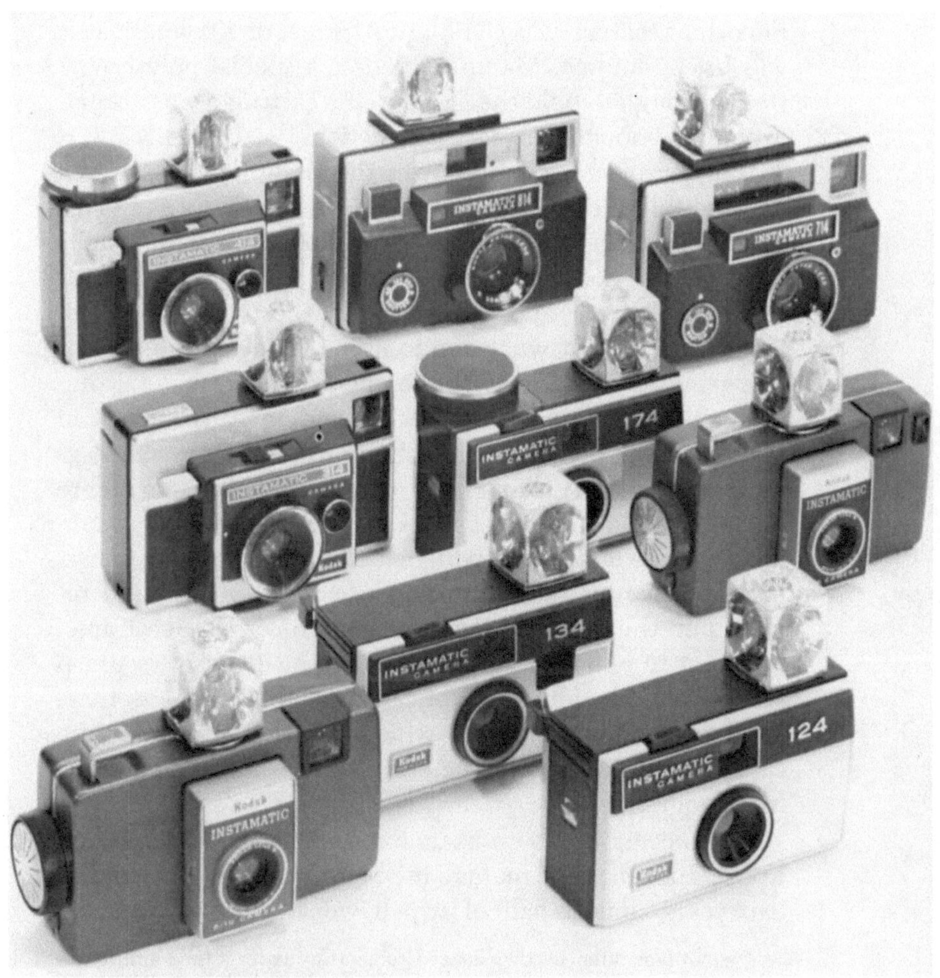

Nine of the reasons why Kodak has been successful in competing in the world market is this lineup of Instamatic cameras. Without automation, acrylic-lens technology, and Yankee ingenuity there would probably be no American camera industry today.

important feature was the film cartridge, or Kodapak, as it was briefly known. Manufactured of a special polystyrene stable enough to hold the film flat, the cartridge was also inexpensive enough for mass production.

When Project 13 was still in the design stages, it was decided to build the concept around a square format. This would utilize the full covering power of a lens with a short focal length, produce a slimmer camera without the need for a collapsible front, and provide more exposures on a shorter length of film. But what film to use? Obviously the answer was 35mm. By this means the company could utilize existing production facilities for that size, and the advantages in photofinishing and slide projection (using the same 2 × 2-inch exterior dimension of the 35mm and 828 frame) were undeniable. The sprocket holes needed to transport 35mm film could be done away with on one edge of the film, allowing expansion of the picture area to 40 × 40mm, and replaced on the other side with a single hole for each frame, sufficient to allow the rapid-transport mechanism to grasp and position the film automatically for the next exposure. A paper backing was provided which allowed a window in the camera back to identify the film in use while protecting it from fogging.[2]

The success was both instant and major. Over 7½ million Kodak Instamatic cameras (in seven models) were sold in two years, almost half of which went overseas.[3] Surveys

[2] 35mm fans often set their loaded cameras aside for a time and forgot what emulsion they were using, a lapse of memory the cartridge user would never suffer.

[3] The stampede is still on and now ranges around the world. In 1969 over 1 million Kodak Instamatic cameras, designed and manufactured in Kodak's plant in Hertfordshire, England, were exported—over and above domestic sales. Kenneth Grange was styling consultant for this line. Kodak announced its 50-millionth Instamatic camera.

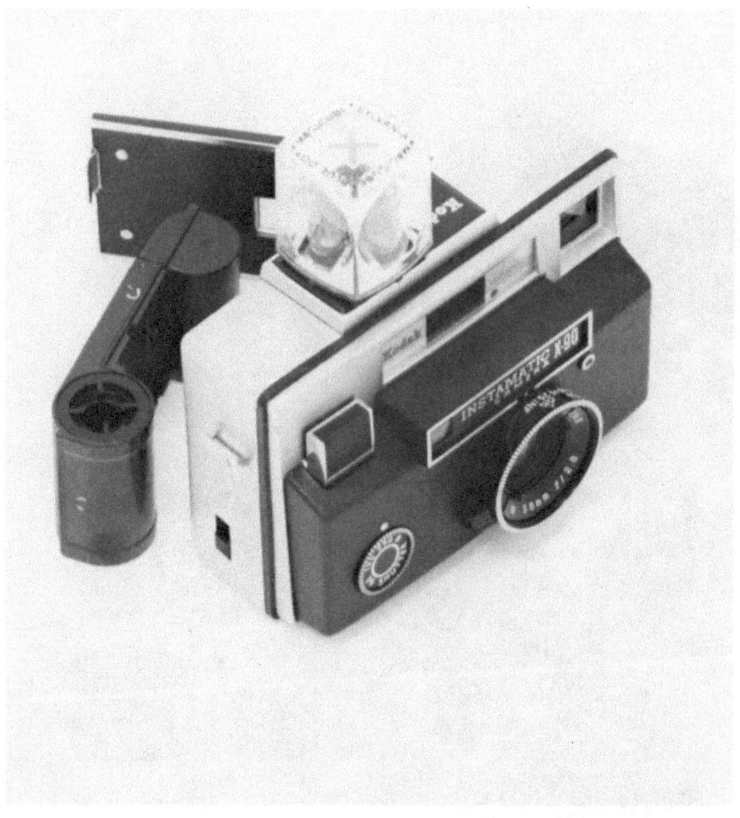

The latest in the long line of top-quality cameras produced in this country and the only one still made here is the Kodak Instamatic X-90. With an Ektar f/2.8 lens and a shutter which is a miniature computer that can override its somewhat limited shutter-speed range, the X-90 is a far more versatile picture maker than some advanced amateurs give it credit for being. But that does not bother volume-conscious Kodak too much, since the X-90 outsells all the rest of the world's quality rangefinder cameras combined.

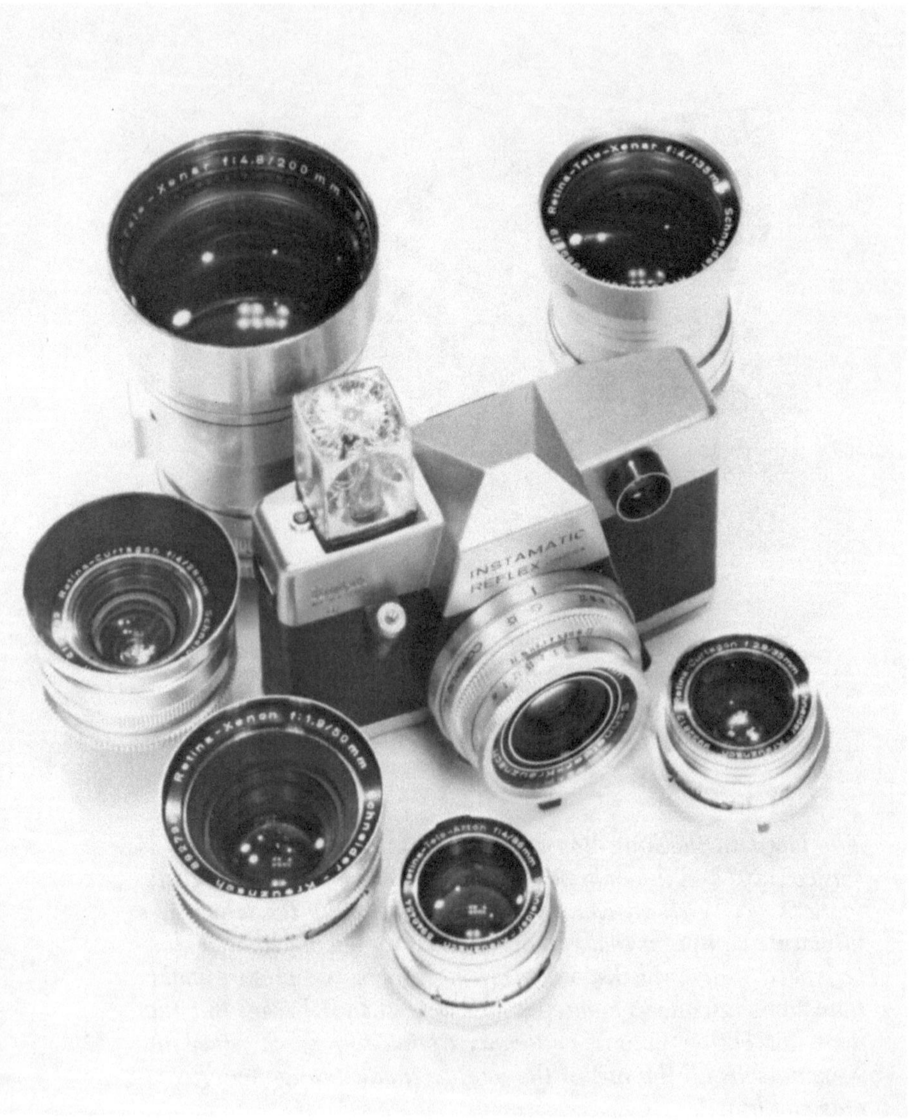
INSTAMATIC
REFLEX
Retina-Xenon f:1.9/50mm
Schneider - Kreuznach

taken at the time indicated that while owners of other cameras used an average of four rolls of film a year Kodak Instamatic owners used eight rolls. In the first 21 months after the introduction of the Instamatic camera, Kodak produced and sold 50 million cartridges of film, which, after all, was the primary goal it had set out to achieve—increased film sales.

German and Japanese manufacturers were caught badly off guard, both by the introduction of the Instamatic and by its overwhelming international acceptance. Led by Agfa-Gevaert, Europe's largest producer of photo goods, the industries of the two countries combined to back the Rapid system as a competitive move, but this 35mm cartridge-to-cartridge concept failed to capture the public's imagination, and, while production of Rapid-loading cameras was immediate and massive, imports to the United States fell drastically for two successive years. As a result, most manufac-

Times have changed! The 35mm rangefinder camera has been gradually replaced in public favor by the single-lens reflex, and today no American manufacturer produces a camera using 35mm film. The instant-loading format has scored very heavily with the amateur photographer. Kodak was the first to supply a top-quality camera using this format, and its Instamatic Reflex replaced the famed Retina line. Its main feature is a unique electronic shutter which is its best and, at the same time, its most limiting feature. The meter is beside rather than through the lens. With the Instamatic Reflex Microscope Adapter ($99) the camera meter in effect reads through the lens by means of a beam-splitting prism system which reflects the light coming from the microscope into the meter. Exposures of up to about 20 seconds are automatically made by the camera shutter. Like the Retina, the Instamatic Reflex is made in Germany by Kodak A.G.

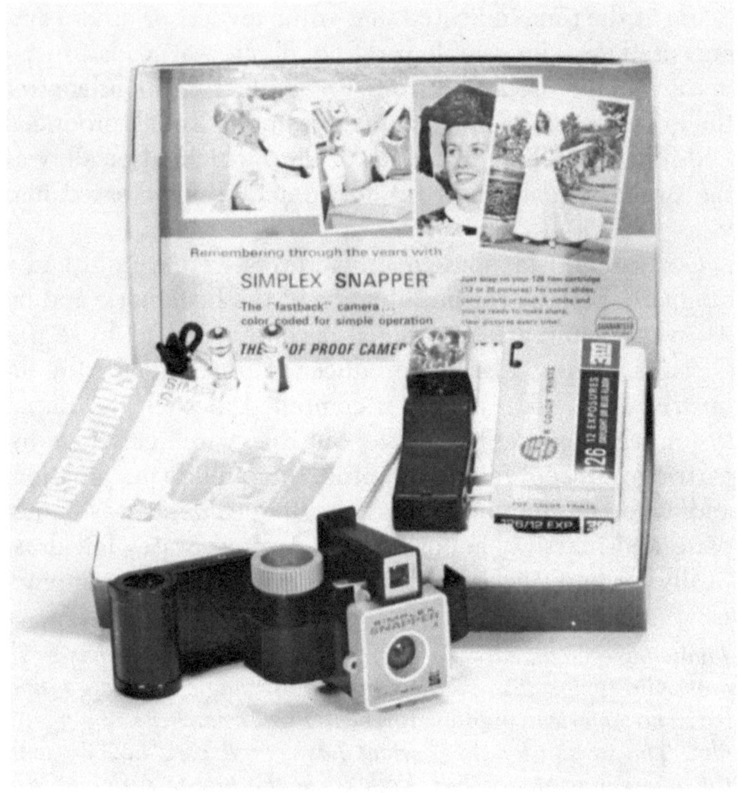

Simply attach a 126 cartridge to this lens and shutter housing, and presto—instant camera. Does this mean that the Simplex Snapper is the ultimate camera design, a camera of the future, a potential throwaway, or perhaps just another in a long line of photographic items that never quite worked out? While the item has great appeal as a novelty, the manufacturer assures purchasers that it is not a toy. Priced at $8.95, it competes with established cartridge cameras, but if the past is really prologue, it may well go the way of mail-in 35mm's—oblivion.

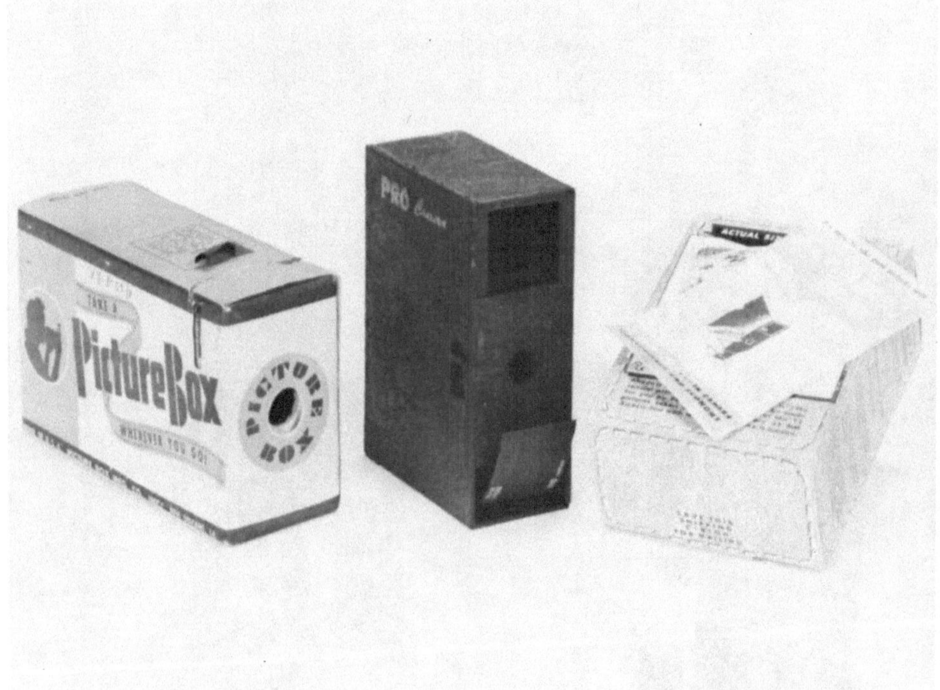

Manufacture of mail-in and disposable cameras began early in the development of photography. Perhaps the ultimate peak was reached with the 35mm market, because the film was inexpensive and the cameras were small. These two examples have deluxe features, including parallax-corrected viewfinders. The Pro (center) was well made of plastic and took 12 pictures. The camera was shipped to the manufacturer in the mailer shown at the right. Neither camera sold well, and both projects were abandoned. Apparently customers did not think the cameras could take good pictures, or maybe the idea of throwing a camera away did not appeal to them.

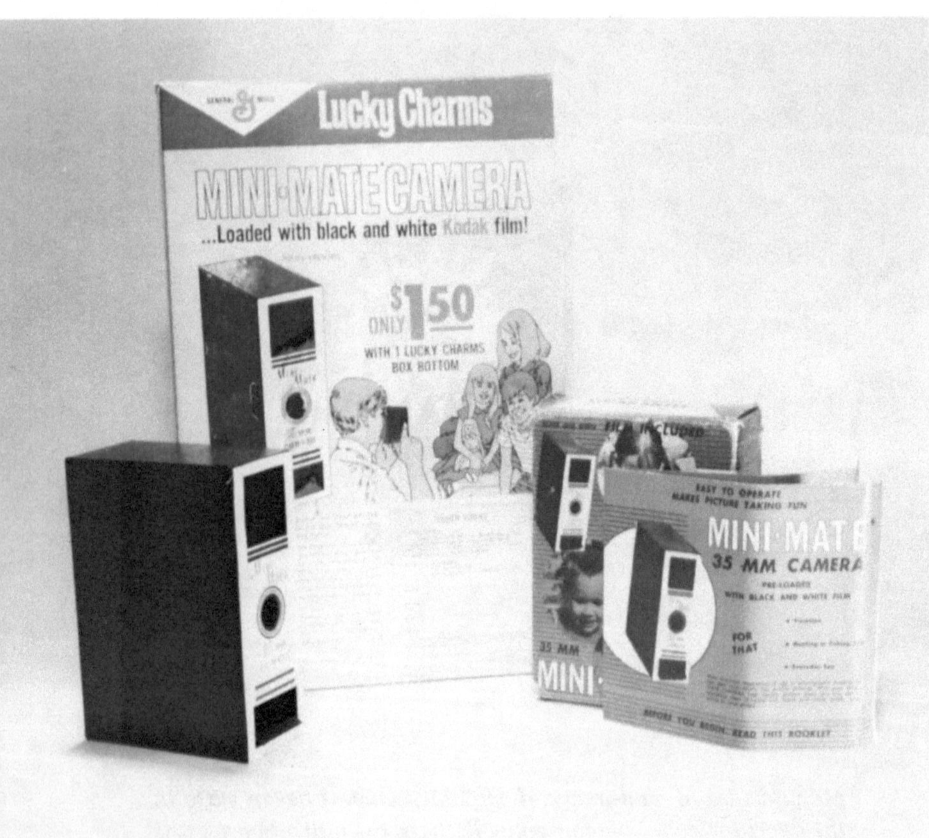

The very last of the American 35mm cameras, the Mini-Mate was hardly a contender for serious photography, but its concept reverted back to that of George Eastman's No. 1 Kodak. The entire camera was mailed to Universal Graphics Corporation, where the film's 12 exposures were processed, the camera was reloaded with film, and was returned complete with prints to the user. Cameras like this were among the more interesting premiums offered to consumers in the 1950's and early 1960's.

turers using the Rapid system have now been licensed to manufacture cameras using 126 film cartridges, and there are still warehouses filled with useless Rapid cameras numbering in the thousands.[4] The Japanese photographic industry was especially hard hit and survived only by forming a cartel to restrict production during 1965–66.

Through it all the 126 Kodak Instamatic cartridge has reigned supreme. Well over 100 different cameras using the 126 cartridge are now on the market, and the number increases each year.[5] Within four years after its introduction, the Kodak Instamatic concept had cut total 35mm sales almost in half (from 600,000 to 325,000), and while 1971 sales figures showed the 35mm camera holding its own, it stands no chance of ever catching up to its brother with the plastic cartridge. The 35mm cameras once manufactured in America died and are now half-forgotten, but the rectangular negative took on new life in a square shape and is firmly established today as the format of the future.

And so we have come to the end of an exciting and creative era which began with the tinkering of Oscar Barnack over half a century ago. What development comes next? Perhaps it will be the noncamera camera, such as the Snapper ($8.95) or the Impro Slip-on (a premium item priced at

[4] Instead of threading the film onto a fixed take-up spool and then having to rewind after the film was used, the Rapid system allowed the user to thread directly into another cartridge, eliminating rewinding. The used cartridge, now empty, became the new take-up cartridge. The system was an outgrowth of the Memo I system and of the earlier Contax magazine system.

[5] The highly sophisticated German-made Kodak Instamatic Single Lens Reflex camera conclusively proves that the 126 cartridge is capable of extremely close tolerances even though it is mass-produced—a sin for which the format has been condemned in the past by purists.

$1.00).[6] Each is nothing more than a lens and a shutter which slip over a 126 cartridge, forming a camera that has the film cartridge as its body. Who knows? It just might catch on.

[6] The Snapper is made for Lava Simplex International, Inc., of Chicago, by Whitehouse Products Company; the Impro, by Impro Camera Corporation, of New York.

Appendix

Camera and Product Manufacturers[1]

Cameras

Argus, Inc. See International Research Corporation
Ansco, Binghamton, N.Y.
Bell & Howell Company, Chicago, Ill.
Bolsey Corporation of America, New York City, N.Y.
Busch Precision Camera Corporation, Chicago, Ill.
Candid Camera Corporation, Chicago, Ill.
Ciro Cameras, Inc., Delaware, Ohio
Clarus Camera Manufacturing Company, Minneapolis, Minn.
Craftsmen's Guild, Hollywood, Calif.
David White Instrument Company, Milwaukee, Wis.
Detrola Corporation, Detroit, Mich.
Eastman Kodak Company, Rochester, N.Y.
Electronics Products Manufacturing Corporation, Ann Arbor, Mich.
Folmer Graflex Corporation, Rochester, N.Y.
International Photographic Industries, New York City, N.Y.
International Research Corporation (later Argus, Inc.), Ann Arbor, Mich.
Keystone Manufacturing Company, Boston, Mass.
Millfred Manufacturing Company, Los Angeles, Calif.
New Ideas Manufacturing Company, New York City, N.Y.
Norton Company, Lockport, N.Y.

[1] The cities given are the locales of the companies during the era of the 35mm camera.

Photographic Industries of America, New York City, N.Y.
Premier Instrument Corporation, New York City, N.Y.
Realist, Inc., Menomonee Falls, Wis.
Revere Camera Corporation, Chicago, Ill.
Simplex Photo Products Company, New York City, N.Y.
Spartus Camera Corporation, Chicago, Ill.
Stereovision, Claude, Texas
Three Dimension Corporation, New Holstein, Wis.
Triad Corporation, Glendale, Calif.
Universal Camera Corporation, New York City, N.Y.
Utility Manufacturing Company, New York City, N.Y.
Vokar Corporation, Dexter, Mich.
Webster Industries, Rochester, N.Y.
Wollensak Optical Company, Rochester, N.Y.

Darkroom Equipment

Burke and James, Inc., Chicago, Ill.
Burleigh Brooks, Inc., New York City, N.Y.
Charles Beseler Company, East Orange, N.J.
Eastman Kodak Company, Rochester, N.Y.
Elwood Pattern Works, Indianapolis, Ind.
Federal Stamping and Engineering Corporation, Brooklyn, N.Y.
F-R Company, Inc., New York City, N.Y.
Ideal, New York City, N.Y.
International Research Corporation, Ann Arbor, Mich.
Kaler Company, Rochester, N.Y.
Kemp Camera Supply, Alhambra, Calif.
Leonard Westphalen, Chicago, Ill.
E. Leitz, Inc., New York City, N.Y.
Simmon Brothers, Long Island City, N.Y.
Sun Ray Photo Company, Inc., New York City, N.Y.
Testrite Company, Newark, N.J.

Optics

Bausch & Lomb Optical Company, Rochester, N.Y.
Elgeet Optical Company, Rochester, N.Y.
General Scientific Corporation, Chicago, Ill.
Graf Optical Company, Chicago, Ill.
Ilex Optical Company, Rochester, N.Y.
Wollensak Optical Company, Rochester, N.Y.

Exposure Devices

Ansco, Binghamton, N.Y.
De Jur Amsco Corporation, Shelton, Conn.
General Electric Company, Schenectady, N.Y.
Intercontinental Marketing Corporation, New York City, N.Y.
Photo Utilities, Inc., New York City, N.Y.
Universal Camera Corporation, New York City, N.Y.
Weston Electrical Instrument Corporation, Newark, N.J.

Lighting Equipment

Abbey Corporation, New York City, N.Y.
Eastman Kodak Company, Rochester, N.Y.
General Electric Company, Cleveland, Ohio
Goodspeed, Inc., New York City, N.Y.
Heiland Research Corporation, Denver, Colo.
Irving Manufacturing Company, Hollywood, Calif.
Kalart Company, New York City, N.Y.
Mendelsohn, S., New York City, N.Y.
J. H. Smith & Sons, Griffith, Ind.
Sylvania Electric Products, Winchester, Ky.
Wabash Photolamp Corporation, Brooklyn, N.Y.
Westinghouse Electric, Pittsburgh, Pa.

Chemicals

Albert Specialty Company, Chicago, Ill. (Micrograin 85)
American Scientific Products, Hollywood, Calif. (D&F)
Chemical Supply Company, Hollywood, Calif. (Champlin)
Eastman Kodak Company, Rochester, N.Y.
Edwal Laboratories, Chicago, Ill.
E. I. du Pont de Nemours & Company, Inc., Wilmington, Del. (Defender)
F-R Company, Inc., New York City, N.Y.
John G. Marshall, Brooklyn, N.Y.
Intercontinental Marketing Corporation, New York City, N.Y. (Perutz)
Oxford Products Company, Beverly Hills, Calif. (Gamma D)

Sensitized Products

Ansco, Binghamton, N.Y.
Dufaycolor, Inc., New York City, N.Y.
Du Pont Film Manufacturing Corporation, New York City, N.Y.
Eastman Kodak Company, Rochester, N.Y.
Gevaert Company of America, Inc., New York City, N.Y.
Intercontinental Marketing Corporation, New York City, N.Y.
Universal Camera Corporation, New York City, N.Y.

Projectors

Eastman Kodak Company, Rochester, N.Y.
Golde Manufacturing Company, Chicago, Ill.
International Research Corporation, Ann Arbor, Mich.
Keystone Manufacturing Company, Boston, Mass.
La Belle Industries, Oconomowoc, Wis.
Spencer Lens Company, Buffalo, N.Y.
Society for Visual Education (SVE), Chicago, Ill.
Three Dimension Corporation, New Holstein, Wis.
Vokar Corporation, Dexter, Mich.

General Index

Index of Cameras

www.ingramcontent.com/pod-product-compliance
Lightning Source LLC
LaVergne TN
LVHW040826090826
845145LV00001BA/207

* 9 7 8 0 8 0 6 1 3 4 3 4 5 *